Brazilian Marine Biodiversity

Series editor
Alexander Turra
São Paulo, Brazil

The book series Brazilian Marine Biodiversity was designed to communicate to a broad and international readership the diversified marine and coastal habitats along the large Brazilian coast.

The diversity of marine habitats found in Brazil is astonishing and includes estuaries, coral reefs, rocky shores, sandy beaches, rhodolith beds, mangroves, salt marshes, deep-sea habitats, vegetated bottoms, and continental shelf. These habitats are addressed from an ecosystem perspective across the series, and characterized in terms of distribution and peculiarities along the Brazilian coast, records of relevant species, and information on the prevailing structuring ecological and oceanographic processes governing biodiversity.

The series also presents an analysis of the role of biodiversity and the importance of ecosystem services, and discusses the threats to each habitat, such as pollution, habitat loss, invasive species, overfishing, and global environmental changes. Conservation efforts are also considered as well as gaps in scientific knowledge and science-policy interface.

This series is an initiative of the Brazilian Network for Monitoring Coastal Benthic Habitats (ReBentos; rebentos.org), which is supported by the Brazilian National Council for Scientific and Technological Development (CNPq), the Research Program on Biodiversity Characterization, Conservation, Restoration and Sustainable Use of the São Paulo Research Foundation (BIOTA-FAPESP), the Coordination for the Improvement of Higher Education Personnel (CAPES) and the Brazilian Innovation Agency (FINEP). ReBentos is part of the Brazilian Network on Global Climate Change Research (Rede Clima) and the Science and Technology National Institute on Climate Changes (INCT Mudanças Climáticas) at the Ministry of Science, Technology, Innovation and Communication (MCTIC).

More information about this series at http://www.springer.com/series/15050

Natalia Pirani Ghilardi-Lopes
Flavio Augusto de Souza Berchez
Editors

Coastal and Marine Environmental Education

Springer

Editors
Natalia Pirani Ghilardi-Lopes
Center for Natural and Human Sciences
Federal University of ABC (UFABC)
São Bernardo do Campo, São Paulo, Brazil

Flavio Augusto de Souza Berchez
Biosciences Institute and CienTec/USP
University of São Paulo (USP)
São Paulo, São Paulo, Brazil

ISSN 2520-1077 ISSN 2520-1085 (electronic)
Brazilian Marine Biodiversity
ISBN 978-3-030-05137-2 ISBN 978-3-030-05138-9 (eBook)
https://doi.org/10.1007/978-3-030-05138-9

Library of Congress Control Number: 2019930143

Cover illustration: Rhodolit close-up - *João Paulo Krajewski*; Fish school at a coral reef, Northeastern Brazil - *Ruy Kenji P. Kikuchi*; Rocky shore intertidal and subtidal fauna, Southeastern Brazil - *Áthila Bertoncini*; Canoes in the estuary of Jaguaribe River, Northeastern Brazil - *Francisco Carlos Rocha de Barros Junior*; Paranaguá estuary, Southern Brazil - *Gisele Morais*; Mangrove at Iriri River, Bertioga, Southeastern Brazil - *Guilherme Moraes de Oliveira Abuchahla*

This Springer imprint is published by the registered company Springer Nature Switzerland AG
The registered company address is: Gewerbestrasse 11, 6330 Cham, Switzerland

Coordination

Support

Foreword

The volume on *Coastal and Marine Environmental Education* (CMEE) represents an updated and broad overview on relevant theoretical and empirical issues that are supporting the development of CMEE in Brazil. This book was elaborated by leading researchers on CMEE in Brazil, from different regions, backgrounds, and institutions, who face the challenge of improving the society's knowledge on the ocean and of creating the conditions for its sustainability. It builds on the rich biodiversity along the Brazilian coast as subsidy to improve the understanding of the society on the ocean. The development of Ocean Literacy is then considered as a relevant milestone to widen and deepen the Environmental Education (EE) approaches, thus aiming to foster changes in the way modern society relates to marine and coastal environments. This is especially relevant given the growing impacts that reach these environments and the relevance of the ocean to humankind.

The concepts of ocean literacy and Anthropocene are presented in the context of the role social control has on ocean environmental quality, followed by how EE initiatives can help in the dissemination of this knowledge. The Brazilian EE setting is then presented focusing on the legal framework. New approaches involving the integration of developing concepts and technologies and the coupling of geo-biodiversity are described, intending to promote the interest of students in environmental issues through an integrated and systemic worldview. Relevant aspects related to the process of evaluation of the effectiveness of EE activities are discussed as a way to improve practices and their achievements, as well as comparisons among different initiatives. The strategic role of Brazilian Marine Protected Areas in promoting both environmental education and conservation is discussed. A synthesis of CMEE projects conducted in Brazil as well as instructional materials used is also presented. This book, thus, comprises a synthesis of the Brazilian conceptual and practical approaches on Coastal and Marine Environmental Education.

Oceanographic Institute
University of São Paulo (USP)
São Paulo, Brazil

Alexander Turra

Preface

Brazil is a country with jurisdiction over a huge marine territory defined by the exclusive economic zone and extended continental shelf, in which the area of approximately 4.5 million km^2 is equivalent to half of its continental territory. Brazil is also a megadiverse country (Paknia et al. 2015) with beautiful and accessible marine and coastal environments, providing a vast field for Coastal and Marine Environmental Education (CMEE). The contact with coastal and marine environments allows for a connection of our feelings and emotions with them. This sensorial commitment to nature can help us free ourselves from the artificial needs of modern societies (Freire 1979; Rousseau 1992). Besides, the contact with and the knowledge regarding the sea and coasts provide people the opportunity to critically reflect about the ecological, social, and economic importance of these ecosystems (Ghilardi-Lopes and Berchez 2013). This contact with nature is good not only under a conservation perspective but also for our physical and mental health (Field et al. 2015). These are basically the assumptions for achieving critical environmental education related to marine and coastal environments, in which human beings are recognized as part of the web of social, natural, and cultural relationships, whose link with the environment results from socially constructed historical processes and whose role in problem identification, resolution, and mitigation of environmental problems must be active and based on solid knowledge (Carvalho 2004). Taking these assumptions into account, the present book, as a part of the book series "Brazilian Marine Biodiversity," aims at presenting to the readers how the accumulated knowledge regarding coastal and marine environments is contributing to the development of CMEE policies, projects, and actions in Brazil. The book is divided into two sections: one related to the conceptual and legal bases of CMEE in Brazil and the other related to the ways CMEE is being implemented in the country.

Since Brazil presents a great diversity of cultural, educational, and social realities, CMEE provides endless possibilities for the promotion of "Ocean and Coastal Literacy" for learners of all ages (NOAA 2013). This is fundamental under the present scenario of profound and complex changes in natural systems caused by human actions (see more on Chap. 1), which demand the construction of knowledge under an interdisciplinary and systemic worldview (García Díaz and Rivero 1996; see more on Chap. 3). At the same time, this diversity imposes several challenges to the development of a standardized Environmental Education (EE) national program (see more on Chap. 2), which also needs to emphasize marine and coastal habitats. The coastal zone of the country encompasses 395 cities distributed in 17 coastal states that have a direct contact with the sea, making CMEE initiatives more accessible. However, there are other 5175 cities in the country to which the "Ocean Literacy Framework" should also reach, especially in elementary and high schools. In the Brazilian curriculum standards, environmental education should be treated across all disciplines, and a way to reach this goal would be to use emerging themes, such as the ocean or climate change, and pedagogies and methodologies, such as phenomenon-based education and outdoor and active learning strategies, which can be associated with Information and Communications Technology (ICT) (see more on Chap. 4).

Independently of the place where they are implemented or the public to which they are targeted, an important concern of educators should be the evaluation of these activities in terms of educational effectiveness and environmental impacts (see more on Chap. 5). As happens in other places around the world, evaluation is not always included as part of planning, making it difficult to draw conclusions regarding the attainment of the goals and objectives initially proposed and the comparison of results across the country.

Regarding the implementation of CMEE in Brazil, the most recent synthesis was carried out by Pedrini (2010) in the book published in Portuguese entitled *Educação Ambiental Marinha e Costeira no Brasil* (*Marine and Coastal Environmental Education in Brazil*). However, since then, educators from all over the country developed and applied a variety of activities in different contexts and used different strategies (see some examples in Chap. 6). Also, relevant initiatives are taking place at Coastal and Marine Protected Areas (CMPAs), although there are some challenges, such as the lack of personnel and funding (see more on Chap. 7). These activities, and the educational materials produced for their accomplishment (see more on Chap. 8), are usually restricted to the places where they were developed and are not properly shared or divulged, which is also a challenge.

We expect this book provides a broad view of CMEE in Brazil, emphasizing our concern with conceptual robustness and sharing possibilities and challenges that can serve as an inspiration for other countries. Enjoy the reading!

São Bernardo do Campo, São Paulo, Brazil — Natalia Pirani Ghilardi-Lopes
São Paulo, São Paulo, Brazil — Flavio Augusto de Souza Berchez

References

Carvalho ICM (2004) Educação Ambiental Crítica: nomes e endereçamentos da educação. In: Layrargues PP (ed) Identidades da educação ambiental brasileira. Ministério do Meio Ambiente, Brasília, p 13–24

Field SC, Lauzon LL, Meldrum JT (2015) A phenomenology of outdoor education leader experiences. J Exp Educ 39:31–44

Freire P (1979) Educação como prática da liberdade. Paz e Terra, Rio de Janeiro

García Díaz JE, Rivero AG (1996) La transición desde un pensamiento simple hacia otro complejo en el caso de la construcción de nociones ecológicas. Investig Esc 28:37–58

Ghilardi-Lopes NP, Berchez F (2013) Mergulho e educação ambiental. In: Oliveira VM, Pereira Filho GH (eds) Mergulho: uma nova ferramenta educacional. da UFRRJ, Seropédica, p 11–22

NOAA (2013) Ocean literacy: the essential principles and fundamental concepts of ocean sciences for learners of all ages. NOAA

Paknia O, Rajaei Sh H, Koch A (2015) Lack of well-maintained natural history collections and taxonomists in megadiverse developing countries hampers global biodiversity exploration. Org Divers Evol 15:619–629

Pedrini AG (2010) Educação Ambiental Marinha e Costeira no Brasil. EdUERJ, Rio de Janeiro, p 272

Rousseau JJ (1992) Emilio ou da Educação. Bertrand Brasil S.A., Rio de Janeiro

Acknowledgments

This book is dedicated to all those who have contributed to the development of Coastal and Marine Environmental Education (CMEE) in Brazil. The editors thank Prof. Alexander Turra for the research initiative that led to funding of the network which allowed for the accomplishment of this book; ReBentos, the Brazilian Network for Monitoring Coastal Benthic Habitats (rebentos.org), which is supported by the Brazilian National Council for Scientific and Technological Development (CNPq); the Research Program on Biodiversity Characterization, Conservation, Restoration and Sustainable Use of the São Paulo Research Foundation (BIOTA-FAPESP); the Coordination for the Improvement of Higher Education Personnel (CAPES); and the Brazilian Innovation Agency (FINEP). We also deeply thank all authors that contributed to this book. Brazil's research agencies, including federal (CNPq and Coordination for the Improvement of Higher Education Personnel—CAPES), state (FAPs), and private agencies, are also acknowledged for the support of research projects that have led to essential scientific assessments of CMEE in the country.

Contents

Contributors

Renato de Almeida Centro de Ciências Agrárias, Ambientais e Biológicas, Universidade Federal do Recôncavo da Bahia (UFRB), Cruz das Almas, Bahia, Brazil

Denise de La Corte Bacci Núcleo de Apoio à Pesquisa – Geohereditas - Instituto de Geociências, University of São Paulo (USP), São Paulo, São Paulo, Brazil

Juliana Imenis Barradas Center for Natural and Human Sciences, Federal University of ABC (UFABC), São Bernardo do Campo, São Paulo, Brazil

Flavio Augusto de Souza Berchez Biosciences Institute and CienTec/USP, University of São Paulo (USP), São Paulo, São Paulo, Brazil

Christine Laure Marie Bourotte Núcleo de Apoio à Pesquisa – Geohereditas - Instituto de Geociências, University of São Paulo (USP), São Paulo, São Paulo, Brazil

Daniel Shimada Brotto Vegetal Biology Department, Rio de Janeiro State University (UERJ), Maracanã, Rio de Janeiro, Brazil

Maria da Glória Motta Garcia Núcleo de Apoio à Pesquisa – Geohereditas - Instituto de Geociências, University of São Paulo (USP), São Paulo, São Paulo, Brazil

Natalia Pirani Ghilardi-Lopes Center for Natural and Human Sciences, Federal University of ABC (UFABC), São Bernardo do Campo, São Paulo, Brazil

Larissa de Araújo Kawabe Center for Natural and Human Sciences, Federal University of ABC (UFABC), São Bernardo do Campo, São Paulo, Brazil

Laura Pioli Kremer Campus Itajaí, Instituto Federal de Santa Catarina (IFSC), Itajaí, Santa Catarina, Brazil

Camila Pinto Meireles Marine Biology and Coastal Environments Postgraduate Program, Fluminense Federal University (UFF), Niteroi, Rio de Janeiro, Brazil

Interdisciplinary Environmental Study Group (GEIA), Rio de Janeiro State University (UERJ), São Gonçalo, Rio de Janeiro, Brazil

Alexandre de Gusmão Pedrini Vegetal Biology Department, Rio de Janeiro State University (UERJ), Maracanã, Rio de Janeiro, Brazil

Douglas de Souza Pimentel Science Department, Rio de Janeiro State University (UERJ), São Gonçalo, Rio de Janeiro, Brazil

Geography Department, Fluminense Federal University (UFF), Niterói, São Gonçalo, Rio de Janeiro, Brazil

Sabrina Gonçalves Raimundo Biosciences Institute, Botany Department, University of São Paulo (USP), São Paulo, São Paulo, Brazil

Maria de Jesus Robim Forestry Institute of São Paulo (IF), State Environmental Secretary (SMA), São Paulo, São Paulo, Brazil

Vânia Maria Nunes dos Santos Núcleo de Apoio à Pesquisa – Geohereditas - Instituto de Geociências, University of São Paulo (USP), São Paulo, São Paulo, Brazil

Antonio Mauro Saraiva Escola Politécnica, University of São Paulo (USP), São Paulo, São Paulo, Brazil

Rosana Louro Ferreira Silva Biosciences Institute, Zoology Department, University of São Paulo (USP), São Paulo, São Paulo, Brazil

Valeria Rocha França da Silva Projeto Tamar, Fundação Pró-Tamar, Salvador, Bahia, Brazil

Suzana Ursi Biosciences Institute, Botany Department, University of São Paulo (USP), São Paulo, São Paulo, Brazil

About the Editors

Natalia Pirani Ghilardi-Lopes is a biologist with a PhD in Marine Ecology from the University of São Paulo. She is currently an Associate Professor at the Federal University of ABC in Santo André (São Paulo State, Brazil). Her main research interests are related to the development and evaluation of marine and coastal environmental education activities, and citizen science protocols. She also develops studies on the ecology of rocky shore benthic communities.

Flavio Augusto de Souza Berchez is a biologist with a PhD in Marine Ecology from the University of São Paulo (USP). He is currently a Professor of the Biosciences Institute at USP. His main research interests are descriptive ecology of hard bottom benthic communities, particularly the habitat patterns and the effects of water motion extreme events, and the creation, application and evaluation of marine ecosystems' environmental education models.

Abbreviations

A3P	Public Administration Environmental Agenda
BA	Bahia State
BIOMAR Network	Network of Projects on Marine Biodiversity
BU	Bibliographic Units
CBME	Community-Based Marine Ecotourism
CDB	The Convention on Biological Diversity
CG	Management Council
CIRM	Interministerial Commission for the Resources of the Sea
CMEE	Coastal and Marine Environmental Education
CMPAs	Coastal and Marine Protected Areas
Conabio	National Biodiversity Commission
CONEMA	State Council of the Environment
COP	Conference of the Parties
CPDS	Commission on Sustainable Development Policies and on the Brazilian Agenda 21
CS	Citizen Science
EE	Environmental Education
EECC	Environmental Education Coordination Committee
EESD	Environmental Education for Sustainable Development
EESS	Environmental Education for Sustainable Societies and Global Responsibility
EEU	Environmental Education Unit
EI	Environmental Interpretation
ENCEA	National Strategy for Communication and Environmental Education in Conservation Units
ESS	Earth System Science
FF	Forest Foundation
FN	Fernando de Noronha
FNDCT	National Fund for Scientific and Technological Development
GCC	Global Climate Change
GEIA	Interdisciplinary Environmental Study Group

IB-USP	University of São Paulo Biosciences Institute
ICMBio	Chico Mendes Institute for Biodiversity Conservation
ICT	Information and Communications Technology
IFSC	Federal Institute of Santa Catarina
IPCC	Intergovernmental Panel on Climate Change
IT	Interpretive Trails
IUCN	International Union for Conservation of Nature
MAI	Itaipu Archaeological Museum
ME	Marine Ecotourism
MEC	Brazilian Ministry of Education
MMA	Brazilian Ministry of the Environment
MPAAB	Marine Protected Area of Armação de Búzios
MPAs	Marine Protected Areas
NFC	Near-Field Communication
PA	Paraíba Aquarium
PAs	Protected Areas
PCNs	Brazilian National Curriculum Parameters
PCNMar	National Policy for the Conservation and Sustainable Use of the Brazilian Marine Biome
PNPCT	National Policy for the Sustainable Development of Traditional People and Communities
PEIA	Anchieta Island State Park
PESET	Serra da Tiririca State Park
PLACEA	Latin American and Caribbean Program of Environmental Education
PNBio	National Policy on Biodiversity
PNEA	National Environmental Education Policy
PNGC	National Coastal Management Plan
PNMA	National Environmental Policy
PNMA I and II	National Environmental Program I and II
PNRH	National Policy on Water Resources
PNRM	National Policy for the Resources of the Sea
PPs	Public Policies
Probio	Project for the Conservation and Sustainable Use of Brazilian Biological Diversity
PROMAR	Marine Mentality Program
Pronabio	National Biological Diversity Program
ProNEA	National Environmental Education Program
ProNMA	National Environmental Program
PSRM	Sectorial Plan for the Resources of the Sea
REARJ	Rio de Janeiro Environmental Education Network
REBEA	Brazilian Environmental Education Network
ReBentos	Monitoring Network for Coastal Benthic Habitats
RESEX Itaipu	Itaipu Marine Extractive Reserve
RFID	Radio-Frequency Identification

S, T & I	Science, Technology, and Innovation
SNUC	National System of Nature Conservation Units
SP	São Paulo State
SPop	Scientific Popularization
TEEE	Transformative and Emancipatory Environmental Education
TEESS	Treaty on Environmental Education for Sustainable Societies and Global Responsibility
TNC	The Nature Conservancy
UC	Nature Conservation Units
UN	United Nations
UNESCO	United Nations Educational, Scientific and Cultural Organization
WWF	World Wildlife Fund

Part I
Conceptual and Legal Bases

Chapter 1
The Importance of "Ocean Literacy" in the Anthropocene and How Environmental Education Can Help in Its Promotion

Natalia Pirani Ghilardi-Lopes, Laura Pioli Kremer, and Juliana Imenis Barradas

Abstract The next generations will face new challenges related to the effects of anthropic impact on the planet. The knowledge about marine socioecological systems will be crucial if we want to maintain a safe operating space for human activities. In this chapter, the concept of "ocean literacy" and its use as a tool to empower the society to cope with this new situation in our history on Earth will be addressed, followed by a brief view of its use on Europe, North America, and South America. After that, the reasons why modern citizens should know, understand, and care about the ocean will be presented. Finally, we will discuss possibilities for the promotion of ocean literacy through environmental education (EE), based on the EE principles established since 1970s.

Keywords Sustainable development goals · Ocean literacy · Anthropocene · Planetary boundaries · Threats to marine ecosystems · Environmental education

1.1 What Is "Ocean Literacy"?

We can generically define the term "ocean literacy" as *the understanding of the ocean's influence on you as well as your influence on the ocean* (West 2004). An ocean literate person can be defined as someone who *understands the fundamental concepts about the functioning of the ocean, can communicate about the ocean in a*

N. P. Ghilardi-Lopes (✉) · J. I. Barradas
Center for Natural and Human Sciences, Federal University of ABC (UFABC), São Bernardo do Campo, São Paulo, Brazil
e-mail: natalia.lopes@ufabc.edu.br

L. P. Kremer
Campus Itajaí, Instituto Federal de Santa Catarina (IFSC), Itajaí, Santa Catarina, Brazil
e-mail: laura.kremer@ifsc.edu.br

N. P. Ghilardi-Lopes, F. A. S. Berchez (eds.), *Coastal and Marine Environmental Education*, Brazilian Marine Biodiversity,
https://doi.org/10.1007/978-3-030-05138-9_1

meaningful way, and is able to make informed and responsible decisions regarding the ocean and its resources (Cava et al. 2005, p. 5). This means that "ocean literacy" is much more than an educational issue; it's a paradigm shift that will result in a positive human behavioral change (Uyarra and Borja 2016).

In the USA, the concern with the urgent need to promote ocean literacy in schools resulted in an effort initiated in 2002 by the College of Exploration and National Geographic Society (NGS) through an online conference (Schoedinger et al. 2010) whose product was the precursor to the work summarized in the ocean literacy principles (seven essential principles and 44 fundamental concepts), as follows (Cava et al. 2005, p. 9):

1. The Earth has one big ocean with many features.
2. The ocean and life in the ocean shape the features of the Earth.
3. The ocean is a major influence on weather and climate.
4. The ocean makes the Earth habitable.
5. The ocean supports a great diversity of life and ecosystems.
6. The ocean and humans are inextricably interconnected.
7. The ocean is largely unexplored.

After the definition of the essential principles, a huge effort was put into the organization of the scope and sequence for grades K-12, until its publication on 2009. The two publications (*Ocean Literacy: The Essential Principles of Ocean Sciences K-12* and the complementary *Ocean Literacy Scope and Sequence for Grades K-12*) make up the ocean literacy framework (Schoedinger et al. 2010).

In Europe, the first conference on ocean literacy was held in 2012 and resulted in a cooperation between European Union, the USA, and Canada on Ocean Literacy and Atlantic Ocean (Tuddenham et al. 2013), demonstrating a concern on this issue amongst several countries. Other ocean literacy efforts in Japan, Portugal, Chile, China, Australia, Canada, and South Africa are in various stages of development (Copejans and Seys 2012).

Considering South American countries, ocean literacy was inserted as a goal in the national policy on ocean in Chile (2017). In Brazil, the Interministerial Commission for the Resources of the Sea (CIRM) aims to stimulate, by means of planned, objective, and continuous actions, the development of a maritime mentality in the Brazilian population, in line with national interests and aimed at a better knowledge of the sea and its resources, of its importance for Brazil, of the responsibility of its sustainable exploitation, and awareness of the need to preserve it (Brasil 2018). However, the efforts of the scientific community and educators related to the promotion of ocean literacy at schools in Brazil, and possibly at other countries in South America, are still punctual and centered on the individual interests of teachers or scientists who are concerned about environmental issues involving the oceans (see examples on Chaps. 4, 6, and 8). In this context, it is evident that South American countries should initiate the organization of a network to foster ocean literacy in national curricular guidelines, so that it can be properly worked in all schools and other educational spaces.

1.2 Why Should a Person Be Ocean Literate?

Uyarra and Borja (2016) state that the importance in talking about ocean literacy relies on the fact that most relevant social groups as policy makers, decision makers, and politicians don’t have enough knowledge about the ocean to make right and important decisions about this environment, because they are not “ocean literate” enough. Similarly, low levels of ocean literacy can be a barrier for citizens to engage in environmentally responsible behavior or consider ocean-related careers (Guest et al. 2015). This is serious if we consider the fact that the ocean is under several threats and continuous changing, especially in the last decades.

1.2.1 The Big Anthropogenic Threats to the Earth in the Last Few Decades (Anthropocene)

It is possible to notice that mankind modifications on ocean are profound, in large part irreversible and will alter the planet conditions for many centuries (Straub et al. 2016). Because of the scientific perception of the unique effects caused by human activities on the planet, an epoch named “Anthropocene” has been suggested to be considered as the third epoch of the Quaternary Period on geological timescale (Stromberg 2013). This conceptual framework, formally introduced in the first decade of the twenty-first century, recognizes the profound imprint of humankind on Earth, proposing that the planet is moving out of Holocene geological epoch. However, only in the first decade of the twenty-first century the term Anthropocene was formally introduced (Crutzen 2006; Crutzen 2002; Steffen et al. 2011). According to this proposal, the impact of humankind on Earth has started in the last part of eighteenth century, when the first evidence of alteration of levels of carbon dioxide and methane in atmosphere caused by human activities was detected from glacial ice data (Crutzen 2006). However, these modifications have undergone an intense growth since 1950, known as the great acceleration phenomenon, and this date has been proposed to formally start the Anthropocene period (Steffen et al. 2011). Indeed, such drastic changes have been evidenced through different data. Over the last few decades, for instance, the population has grown from around 2.5 billion in 1950 to 7.3 billion today (Engelman and Terefe 2014). Moreover, current species extinction rates, influenced by human activities, are estimated to be tens to hundreds of times higher than natural background rates of extinction. In addition, the concentration of atmospheric carbon dioxide has risen around 26.2% from 1950 to 2013, which is astonishing when compared with the increase of 12.4% in the 200 preceding years (1750–1950) (Steffen et al. 2015; Gaffney and Steffen 2017).

Taking all these changes caused by the interference of humankind on Earth into account, the planetary boundaries (PB) approach was proposed aiming at to define the biophysical limits inherent to the functioning of Earth to maintain a stable state of systems condition for global sustainability and human welfare (Rockström et al. 2009a, b; Steffen et al. 2015). According to this proposal, "the term planetary boundaries has been introduced to define and quantify parameters derived from careful scientific measurement and analysis of the state of earth systems that define a safe operating space for humanity" (Rockström et al. 2009a). In this way, nine large environmental processes, in which safe critical boundaries for sustaining the resilience of ecosystems are crucial to be defined, have been highlighted: climate change, ocean acidification, stratospheric ozone depletion, global phosphorus and nitrogen biogeochemical flows, atmospheric aerosol loading, freshwater use, land-system (use) change, biosphere integrity (biodiversity loss), and novel entities (chemical pollution; Steffen et al. 2015). Four of them (climate change, biosphere integrity, land-system change, and biogeochemical flows) have already been crossed from the safe operating space, being the biggest threats for the planet resilience (Fig. 1.1) (Rockström et al. 2009a; Steffen et al. 2015). Although all these boundaries are defined individually, they are all intertwined and crossing the limits of any of them will modify several other variables.

1.2.2 Our Influences on the Oceans

All nine planetary boundaries (Steffen et al. 2015) are directly or indirectly related to the ocean and these interconnections are still not entirely known by scientists (Rockström et al. 2009b). Even though resource managers and decision makers take actions to either avoid or respond to ecosystems changes, there is a possibility that some marine ecosystems undergo large, sudden, and surprising changes in response to multiple stressors, resulting in dramatic shifts in their structure and functioning that are often costly and hard to reverse (tipping points) (Selkoe et al. 2015). These changes can cause great impacts on physical, chemical, and biological processes on continental areas and ocean basins and affect the living conditions and human health and well-being (Sauerborn and Matthies 2006; IPCC 2014). As an example, we can cite industrial chemical pollution, ocean acidification, sewage discharges, oil spills, changes to beaches, shores, and rivers, and overfishing (NOAA 2013). An evaluation of the human impact on the seas reveals that around 40% of the world's ocean has been heavily affected by human activities, including fishing, coastal development, and pollution from shipping (Halpern et al. 2008). This is not surprising, since about half of the world's population lives within 200 km of the coast and around 65% of cities with more than five million inhabitants are located in coastal zones (Nyerges 2014; Rumson and Hallett 2018; Small and Nicholls 2003), imposing several impacts to these areas. Besides, not only the impacts of the cities close to the ocean are relevant, but also the water footprint of the big cities located far from it (van Leeuwen 2013).

Among human impacts on ocean, it is worth noting the ocean acidification (one of the nine planetary boundaries) caused by increasing amount of atmospheric

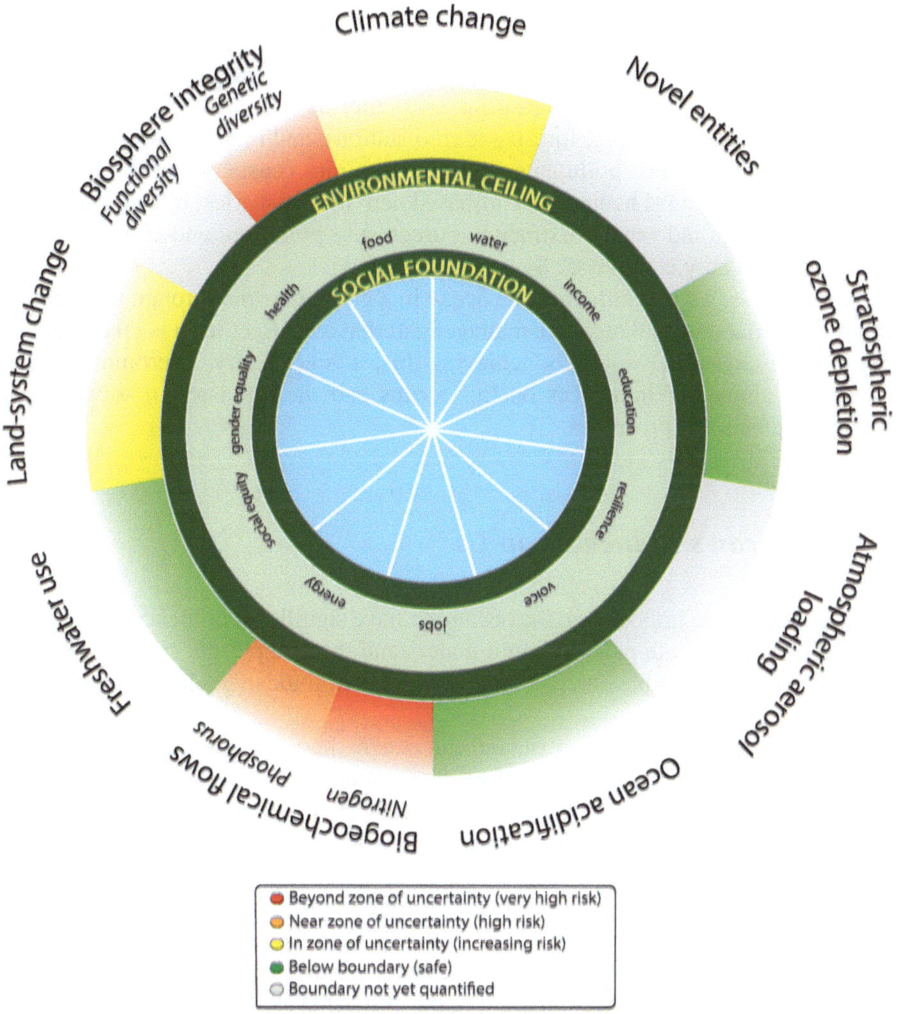

Fig. 1.1 Nine planetary boundaries and their state of risk (see legend for the meaning of different colors). The "doughnut" model represents the safe operating space for humanity, considering the social foundation and the limitation by environmental ceiling (Rockström et al. 2009a, Raworth 2017)

carbon dioxide produced by the burning of fossil fuels since the beginning of the industrial revolution. The carbon dioxide is absorbed by ocean causing a shift in the seawater chemistry, increasing aqueous CO_2 and decreasing pH and carbonate ion (CO_3^{2-}) concentrations. This alteration in water chemistry can affect the saturation states of calcium carbonate minerals such as calcite and aragonite, which are important constituents of shells and skeletons of several organisms (Feely et al. 2004; Orr et al. 2005). In this way, some marine species can be seriously affected, such as mollusks and crustaceans, which are valuable sources for commercial and recreational fisheries (Cooley et al. 2009); and coral reefs, which support a variety of subsistence, recreational, and commercial activities worldwide (Bryant et al. 1998).

Land-based activities like coastal infrastructure development, industrial activities, agriculture/aquaculture, forestry (deforestation), and recreation/tourism affect the runoff of sediments, pollutants, and nutrients into coastal waters and remove, alter, or destroy natural habitat. Ocean-based activities (fisheries, oil and gas exploration, shipping, and others) extract resources, add pollution, and change species composition (Pauly et al. 2005; Syvitski et al. 2005).

Nonetheless, we can influence the ocean in a good way too, through the creation of laws, regulations, and resource management that can affect what is taken out and put into the ocean. However, for achieving this, it is necessary, to promote critical awareness about our influences on the oceans and the influence of ocean on us through educational processes.

1.2.3 Ocean's Influences on Us

All changes we are imposing to the ocean can have significant influence on our daily lives. An ocean literate person should understand that the present conditions of the planet were only possible due to a long history of biological, geological, and physicochemical processes.

Covering about 70% of our planet, the oceans have influenced Earth since its formation, estimated in 4.6 billion of years, what means, since the beginning of our planet (Owen and Walker 1999). Ocean has shaped the surface of land through changes in the sea level over time that have expanded and contracted continental shelves, created and destroyed inland seas (NOAA 2013).

Most rain that falls on land originally evaporated from the tropical ocean, what gives it a key role in the vital cycle of water, being responsible for 86% of global evaporation, which is also important in the heat exchange between the ocean and the atmosphere, playing a central function in the climate system (Bollmann et al. 2010; NASA 2017).

The ocean also acts as a CO_2 reservoir and can absorb ten times more carbon dioxide than freshwater. The carbon is easily dissolved in ocean waters and may be stored in large quantities in the marine sediments. All this carbon is continuously recycled within the planktonic chain (phytoplankton, zooplankton) (Odum 1988; Bollmann et al. 2010).

Over 1.3 billion people, mostly in the developing countries, live in coastal communities bordering tropical seas and many depend on adjacent coastal seas for food and livelihoods (Sale et al. 2014).

The ocean also regulates weather and climate and provides a huge amount of ecosystem services, like food, medicines, mineral, and energy resources, supporting many economies worldwide that depends directly on it (Cooley et al. 2009).

According to Brown et al. (2007), ecosystem services are "*the specific results of ecosystem processes that either directly sustain or enhance human life or maintain the quality of ecosystem goods.*" Ecosystem services are usually divided into four categories: provisioning, regulation, culture, and support and can generate direct income or provide benefits that are more difficult to constrain, such as cultural identity (Millennium Ecosystem Assessment 2005). Some methods have been developed to measure ecosystem services as presented in the work of Burke et al. (2008) and Cooper et al. (2009), which estimate that ecosystem services from the sea in some Caribbean islands are around millions of dollars a year with coastal protection by reefs and mangroves, direct economic benefits, and indirect taxes on tourism and fisheries.

Besides all those things that can be measured, ocean is a source of inspiration, recreation, rejuvenation, and discovery. It is also an important element in the heritage of many cultures (NOAA 2013).

Despite all these benefits, and although we have already mapped the entire ocean floor with a resolution of about 5 km (Witze 2017), we still know very little about the functioning of oceanic ecosystems. Therefore, understanding minimally the complexity of factors that interact within the ocean and among land, atmosphere and ocean is a key to reach ocean sustainability. We are all responsible for the ocean and life on Earth was only possible because of it. More recently, the oceans were explicitly included amongst the 17 sustainable development goals of ONU 2030 agenda for sustainable development (Goal 14: Conserve and sustainably use the oceans, seas, and marine resources for sustainable development) (United Nations, 2015), emphasizing its relevance in the international agenda.

In this context, environmental education can be considered as a process that can foster sensitivity, awareness, and understanding, to empower citizens in making critical decisions related to the ocean, being a key strategy to promote ocean literacy and achieve ocean sustainability (UNESCO 2017).

1.3 How Environmental Education Can Help in Promoting "Ocean Literacy"?

Environmental education is an important process which aims at supplying people with knowledge, sensitivity, awareness, and skills related to socio-environmental issues, which can lead to a motivation and commitment of citizens, influencing their attitudes and promoting actions that help to achieve positive environmental change (Ardoin and Merrick 2013).

Since Tbilisi's Conference on 1977, 12 environmental education (EE) principles were established, and EE activities and programs must consider them. In this context, oceanic literacy is essential for an effective construction of the fundamentals in coastal and marine environmental education (CMEE), as mentioned below for each EE principle.

1.3.1 Consider the Environment in Its Totality: Natural and Built, Technological and Social (Economic, Political, Cultural–Historical, Ethical, Esthetic)

Regarding this principle, aspects related to the ocean can be easily worked by educators, such as ocean biodiversity, urbanization of coasts and construction of infrastructure (harbors and marinas, for example), development of marine transport means or technology (related to tidal energy capture, mining, fossil fuels, among others), the use of the ocean by different social actors (for example, traditional communities, managers, tourists, politicians, educators, and researchers), the laws related to the ocean and their use by different countries, the importance of the ocean as provider of ecosystem services, as climate regulation source of inspiration and well-being.

1.3.2 Be a Continuous Lifelong Process, Beginning at the Preschool Level and Continuing Through all Formal and Non-formal Stages

For the ocean, this principle can be only possible with the development of ocean literacy framework in the curricular guidelines of different countries. In the Brazilian National Common Curricular Basis (BNCC—available at http://basenacionalcomum.mec.gov.br/download-da-bncc) for elementary school years, the oceanic issues are treated in the disciplines of geography (pollution of the ocean) and history (navigations; complexity and interactions of Atlantic, Pacific, and Indian Oceans) (Brasil 2017).

1.3.3 Be Interdisciplinary in its Approach, Drawing on the Specific Content of Each Discipline in Making Possible a Holistic and Balanced Perspective

We can say that knowledge on ocean is interdisciplinary by nature, including, among others: the chemistry of ocean waters; the physics related to currents; the huge oceanic biodiversity and ecological interactions in space and time; the geology of ocean transformations along time; the mathematics in, for example, pH logarithmic scale; the technology involved in maritime transportation or energy production; the economy based on oceanic resources; the political aspects involving exploitation and conservation of oceanic resources and ecosystem services, and the historical facts involving human relationship with ocean (for example, the great navigations and new land discoveries or deprivation of marine resources for food or economic activities). This complexity constitutes a vast field for educators of all disciplines and, at the same time, a possibility for their interaction on the development of environmental education activities focused on oceanic issues.

1.3.4 *Examine Major Environmental Issues from Local, National, Regional, and International Points of View so that Students Receive Insights into Environmental Conditions in Other Geographical Areas*

Considering the planetary boundaries approach, all nine boundaries and their relationship with the ocean could serve as starting points for the discussion of marine issues on an international/global point of view. After that, contextualized examples on local, regional, or national scales can be studied in a deeper perspective and compared to examples on other localities of the planet. As one example, the discussion on how the alteration of ocean currents due to climate change can affect the economy and climate in different parts of the Earth can be stimulated.

1.3.5 *Focus on Current and Potential Environmental Situations While Considering the Historical Perspective*

Some decades ago, especially during the industrial revolution, we didn't have enough knowledge to notice the damages that our activities were causing to the environment in a global scale. It wasn't clear what could happen in the future. Nowadays, we have new technologies and accumulated knowledge that help us better predict and prevent unsustainable actions, although the knowledge about the complexity of natural systems shows us that it is difficult to accurately predict the future (García Díaz and Rivero 1996). Working on issues affecting the oceans today, such as warming and acidification of waters, but contextualizing them in the history of planet Earth and of the humankind could be an important tool to help educators teach about the influence of human activities on nature in recent periods.

1.3.6 *Promote the Value and Necessity of Local, National, and International Cooperation in the Prevention and Solution of Environmental Problems*

The first time this principle was considered was in 1967, when the United Nations' ambassador, Arvid Pardo, in his speech at one of the general assemblies, addressed the need for updating the freedom-of-the-seas doctrine due to the technological changes that had altered men's relationship to the ocean (United Nations 1967, 2012). In the following years, the United Nations Seabed Committee was created, and among the established issues there was a declaration that all resources of the seabed beyond the limits of national jurisdictions should be the common heritage of humankind. It turned into a global effort to regulate and write rules for all ocean areas, all uses of the seas and all their resources. These were some of the factors that led to the convening of the Third United Nations Conference on the Law of the Sea to write a comprehensive treaty for the oceans that ended up 9 years later with the adoption of a constitution for the seas—the United Nations Convention on the Law of the Sea. After that, additional agreements were adopted to further extend the legal regime to assign each country its responsibility over the ocean and seas (United Nations 2012).

The role of EE in this principle is to present these historical events, mostly unknown for most people, to emphasize our individual and collective obligation to protect and manage ocean resources.

1.3.7 Explicitly Consider Environmental Aspects in Plans for Development and Growth

In this principle, EE can help to discuss the impacts that can be caused due to activities related to growth and development. Usually, citizens only know the impacts of an activity after it's already installed and sometimes hardly affecting their lives. When people are aware that the development and growth of a city can bring together possible environmental impacts that surely will affect them, it's possible to delineate a plan for a better solution or even contest the government actions.

1.3.8 Enable Learners to Have a Role in Planning Their Learning Experiences and Provide an Opportunity for Making Decisions and Accepting Their Consequences

Planning the activities someone is about to do is a very good exercise to notice how planning is important and to evaluate our own decisions later. For example, considering the principle 7 listed above, it is possible to apply an activity about the possibility of the construction of a new port in a certain city, as happened recently at Araça Bay in the state of São Paulo in Brazil (Amaral et al. 2010). In groups, the learners can do a research, looking for some other examples of port construction in other places, considering their bad and good aspects and based on that, present to the others their point of view, which can be for or against it.

This kind of discussion can be very productive because it creates the opportunity to think abroad, considering all aspects and considering different points of view. Maybe one group could think about the aspect of water pollution related to the port construction, but another one could think about the traditional communities that depend on fishery (see https://www.youtube.com/watch?v=HtGd2t1YSJ8, for an example related to Araça Bay), in a way that both points of view complement each other.

1.3.9 Relate Environmental Sensitivity, Knowledge, Problem-Solving Skills, and Values Clarification to Every Age, But with Special Emphasis on Environmental Sensitivity to the Learner's Own Community in Early Years

Environmental sensitivity is an important variable in environmental awareness and in the predisposition to take responsible environmental action (Chawla 1998). The ocean presents a wide range of knowledge that could be dealt with in formal education, in an interdisciplinary way, and in informal education (see Tbilisi's EE principles 1 and 3). The ocean literacy framework provides conceptual flow diagrams with concepts, subconcepts, and their approach for scholar curriculum for grades K-12. Moreover, the inextricably interdependence between humankind and ocean and the problems that arise from this interaction enable the learning in a problem-solving perspective. In addition, the coastal ecosystems can awaken several affective attributes, essential to environmental sensitivity, like curiosity, interest, satisfaction, happiness, and fear.

1.3.10 Help Learners Discover the Symptoms and Real Causes of Environmental Problems

The ocean is interdependent with the global system, since it shapes the features of Earth (Ocean Literacy Framework—Principle 2), influences the weather and climate (Ocean Literacy Framework—Principle 3) and makes Earth habitable (Ocean Literacy Framework—Principle 4). In such a way, the ocean literacy is essential for understanding the dynamics of Earth system and it is not possible to discover the causes of environmental problems and make conscious decisions about these without understanding the ocean. In this context, EE activities can help learners identify and investigate environmental problems related to the ocean, such as pollution and overexploitation (in a local/regional scale) or acidification (on a global scale). Conversely, they may also stimulate the learners to identify the effect of the oceans on the local livelihood conditions. Outdoor experiential learning and problem-based learning can help learners understand the deep causes of environmental problems related to ocean and, consequently, promote reflection and discussion on possible solutions.

1.3.11 Emphasize the Complexity of Environmental Problems and Thus the Need to Develop Critical Thinking and Problem-Solving Skills

Critical thinking about how the ocean is interconnected with our lives (see Sect. 1.2 for more information) is essential to think about the interrelationships between individual and collective actions and environmental problems. For instance, the critical understanding of global climate change and renewable energy sources requires complex connections between knowledge related to the burning of fossil fuels, the carbon cycle, and the ocean's ability to store carbon and heat. This way, to capacitate our students to better deal with the problems related to these complex systems, environmental education activities should work on important skills for the citizens of the twenty-first century, such as conflict solving, responsibility, reasoning and argumentation, adaptive learning, critical thinking, cooperation, self-learning, among others (National Research Council 2012).

1.3.12 Utilize Diverse Learning Environments and a Broad Array of Educational Approaches to Teaching, Learning About and from the Environment with Due Stress on Practical Activities and First-Hand Experience

The classroom, laboratories, museums, or marine protected areas can be all locations in which learners can be in touch with the ocean's issues. It is known that childhood outdoor experiences such as free play, hiking, camping, fishing, and berry picking (Chawla and Cushing 2007) can distinguish environmentally active people from those who show less commitment.

At schools, there are simple experiments that can be done to promote the understanding of some ocean processes linking them to curriculum concepts. For example, to explain ocean currents and density concepts an experiment could be performed by adding a dye in hot water or salt water and mixing it with cold freshwater. In addition, there are several media that are related to the ocean, including animations for children, which can be explored in a critical perspective. Moreover, there are recreational and educational materials that could be explored in formal and informal education (see examples on Chap. 8).

1.4 Final Remarks

In the last decades ocean is facing huge alterations, mostly due to anthropogenic actions. However, our knowledge is still limited and oceanic issues are not formally inserted in curriculum guidelines of Brazilian and other South American countries, which can be a barrier for the promotion of ocean literacy. If we want to reach ocean sustainability, ocean literacy framework is fundamental for all ages, but especially for young learners. Due to the natural complexity and interdisciplinarity of oceanic issues, planning and implementing environmental education activities involving the ocean is natural and can help fostering the development of skills necessary for the twenty-first century citizens to face the unpredictable and changing scenarios of the risk society (sensu Beck 1992).

Acknowledgements We thank Peterson Lásaro Lopes for the elaboration of Fig. 1.1.

References

Amaral ACZ, Migotto AE, Turra A, Schaeffer-Novelli Y (2010) Araçá: biodiversidade, impactos e ameaças. Biota Neotropica 10:219–264

Ardoin N, Merrick C (2013) Environmental education: a brief guide for U.S. grantmakers

Beck U (1992) Risk society: towards a new modernity. SAGE, Newbury Park

Bollmann M, Bosch T, Colijn F et al (2010) World ocean review: living with the oceans. Maribus GmbH, Hamburg

Brasil (2017) Base Nacional Comum Curricular: educação é a base. Ministério da Educação, Brasília

Brasil (2018) Programa de Mentalidade Marítima da Comissão Interministerial para os Recursos do Mar (CIRM). Available at: https://www.marinha.mil.br/secirm/promar. Accessed 23 June 2018

Brown TC, Bergstrom JC, Loomis JB (2007) Defining, valuing, and providing ecosystem goods and services. Nat Resour J 47:329–376

Bryant D, Burke L, Mcmanus J, Spalding M (1998) Reefs at risk: a map-based indicator of threats to the world' s coral reefs. World Resources Institute, Washington

Burke L, Prager D, Cooper E, Greenhalgh S (2008) Coastal capital–economic valuation of coral reefs in Tobago and St. Lucia. World Resource Institute, Washington

Cava F, Schoedinger S, Strang C, Tuddenham P (2005) Science content and standards for ocean literacy: a report on ocean literacy. Available at: http://www.coseeca.net/publications/OLit04-05FinalReport.pdf

Chawla L (1998) Significant life experiences revisited: a review of research on sources of environmental sensitivity. J Environ Educ 3:11–21

Chawla L, Cushing DF (2007) Education for strategic environmental behavior. Environ Educ Res 13:437–452

Chile (2017) Política oceánica nacional de Chile. Available at: https://minrel.gob.cl/minrel/site/artic/20180309/asocfile/20180309143025/poli__tica_ocea__nica_nacional_de_chile_ok.pdf. Accessed 19 June 2018

Cooley SR, Kite-Powell H, Doney SC (2009) Ocean acidification' s potential to alter global marine ecosystem services. Oceanography 22:172–181. https://doi.org/10.5670/oceanog.2009.106

Cooper E, Burke L, Bood N (2009) The economic contribution of Belize's coral reefs and mangroves. World Resource Institute, Washington

Copejans E, Seys J (eds.) (2012) First conference on ocean literacy in Europe – book of abstracts. Bruges, Belgium, 12 October 2012. VLIZ special publication 60 – Flanders Marine Institute (VLIZ). Oostende, Belgium

Crutzen PJ (2002) Geology of mankind. Nature 415:23

Crutzen PJ (2006) The "Anthropocene". In: Ehlers E, Krafft T (eds) Earth System Science in the Anthropocene. Springer Berlin Heidelberg, Berlin, Heidelberg, pp 13–18. https://doi.org/10.1007/3-540-26590-2_3

Engelman R, Terefe Y (2014) Will population growth end in this century? Vital signs, Washington, pp 1–7

Feely RA, Sabine CL, Lee K et al (2004) Impact of anthropogenic CO_2 on the $CaCO_3$ system in the oceans. Science 305:362–366. https://doi.org/10.1126/science.1097329

Gaffney O, Steffen W (2017) The Anthropocene equation. Anthr Rev 4:53–61. https://doi.org/10.1177/2053019616688022

García Díaz JE, Rivero AG (1996) La transición desde un pensamiento simple hacia otro complejo en el caso de la construcción de nociones ecológicas. Investigación en la Escuela 28:37–58

Guest H, Lotze HK, Wallace D (2015) Youth and the sea: ocean literacy in Nova Scotia, Canada Mar Policy 58:98–107. https://doi.org/10.1016/j.marpol.2015.04.007

Halpern B, Walbridge S, Selkoe K et al (2008) A global map of human impact on marine ecosystems. Science 319:948–952. https://doi.org/10.1126/science.1151084

IPCC (2014) Climate change 2014 synthesis report summary chapter for policymakers. Available at: https://www.ipcc.ch/site/assets/uploads/2018/02/AR5_SYR_FINAL_SPM.pdf. Accessed 21 january 2019

Millennium Ecosystem Assessment (2005) Ecosystems and human well-being: synthesis. Island Press, Washington

NASA (2017) Water cycle. In: NASA GSFC water energy cycle website. Available at: http://science.nasa.gov/earth-science/oceanography/ocean-earth-system/ocean-water-cycle. Accessed 21 january 2019

National Research Council (2012) Education for life and work: developing transferable knowledge and skills in the 21st century. The National Academies Press, Washington

NOAA (2013) Ocean literacy. The essential principles and fundamental concepts of ocean sciences for learners of all ages (version 2). Available at: http://www.coexploration.org/oceanliteracy/documents/OceanLitChart.pdf. Accessed 21 january 2019

Nyerges T (2014) Introduction to the Special Issue: Coastal GIS. ISPRS Int J Geo-Inf 3(3):1118–1121. https://doi.org/10.3390/ijgi3031118

Odum E (1988) Ecologia. Editora Guanabara Koogan S.A., Rio de Janeiro

Orr JC, Fabry VJ, Aumont O et al (2005) Anthropogenic ocean acidification over the twenty-first century and its impact on calcifying organisms. Nature 437:681–686. https://doi.org/10.1038/nature04095

Owen T, Walker JC (1999) What do we know about the origin of the earth's oceans? Is it more likely that they derive from icy comets that struck the young earth or from material released from the earth's interior during volcanic activity? Sci Am:1–6

Pauly D, Watson R, Alder J (2005) Global trends in world fisheries: impacts on marine ecosystems and food security. Phil Trans R Soc B 360:5–12. https://doi.org/10.1098/rstb.2004.1574

Raworth K (2017) A Doughnut for the Anthropocene: humanity's compass in the 21st century. Lancet Planet Health 1:e48–e49. https://doi.org/10.1016/S2542-5196(17)30028-1

Rockström J, Steffen W, Noone K et al (2009a) A safe operating space for humanity. Nature 461:472–475. https://doi.org/10.1038/461472a

Rockström J, Steffen W, Noone K et al (2009b) Planetary boundaries: exploring the safe operating space for humanity. Ecol Soc 14:32 [online] URL: http://www.ecologyandsociety.org/vol14/iss2/art32/

Rumson AG, Hallett SH (2018) Opening up the coast Ocean & Coastal Management 160:133–145. https://doi.org/10.1016/j.ocecoaman.2018.04.015

Sale PF, Agardy T, Ainsworth CH et al (2014) Transforming management of tropical coastal seas to cope with challenges of the 21st century. Mar Pollut Bull 85:8–23. https://doi.org/10.1016/j.marpolbul.2014.06.005

Sauerborn R, Matthies F (2006) Global Environmental Change and Human Health. In: Ehlers E, Krafft T (eds) Earth System Science in the Anthropocene. Springer Berlin Heidelberg, Berlin, Heidelberg, pp 113–134. https://doi.org/10.1007/3-540-26590-2_10

Schoedinger S, Tran LU, Whitley L (2010) From the principles to the scope and sequence: a brief history of the ocean literacy campaign. NMEA Spec Rep 3:3–7

Selkoe KA, Blenckner T, Caldwell MR et al (2015) Principles for managing marine ecosystems prone to tipping points. Ecosyst Health Sustain 1:art17. https://doi.org/10.1890/EHS14-0024.1

Small C, Nicholls RJ (2003) A Global Analysis of Human Settlement in Coastal Zones. J Coast Res 19:584–599

Steffen W, Grinevald J, Crutzen P, Mcneill J (2011) The Anthropocene: conceptual and historical perspectives. Phil Trans R Soc A 369:842–867. https://doi.org/10.1098/rsta.2010.0327

Steffen W, Richardson K, Rockstrom J et al (2015) Planetary boundaries: guiding human development on a changing planet. Science 347(6223):1259855. https://doi.org/10.1126/science.1259855

Straub SC, Thomsen MS, Wernberg T (2016) The dynamic biogeography of the Anthropocene: the speed of recent range shifts in seaweeds. In: Hu Z-M, Fraser C (eds) Seaweed phylogeography: adaptation and evolution of seaweeds under environmental change. Springer, Dordrecht, pp 63–93. https://doi.org/10.1007/978-94-017-7534-2_3

Stromberg J (2013) What is the anthropocene and are we in it? Smithsonian Magazine. Available at: https://www.smithsonianmag.com/science-nature/what-is-the-anthropocene-and-are-we-in-it-164801414/. Accessed 21 June 2018

Syvitski JPM, Vörösmarty CJ, Kettner AJ, Green P (2005) Impact of humans on the flux of terrestrial sediment to the global coastal ocean. Science 308:376–380. https://doi.org/10.1126/science.1109454

Tuddenham P, Grigorov I, Bishop T, et al (2013) Vision statement on ocean literacy and Atlantic Ocean Cooperation between European Union, United States of America & Canada. In: Transatlantic ocean literacy workshop 2013. Plymouth, pp 1–5

UNESCO (2017) Education for sustainable development goals - learning objectives. Available at: http://www.unesco.org/open-access/terms-use-ccbysa-en. Acessed 21 June 2018

United Nations (1967) United Nations general assembly twenty-second session. Available at: http://www.un.org/Depts/los/convention_agreements/texts/pardo_ga1967.pdf. Accessed 21 january 2019

United Nations (2012) Oceans and law of the seas. Available at: http://www.un.org/Depts/los/convention_agreements/convention_historical_perspective.htm. Accessed 21 january 2019

United Nations (2015) Transforming our world: the 2030 agenda for sustainable development. General assembly 70 session 16301, pp 1–35. Available at: https://sustainabledevelopment.un.org/post2015/transformingourworld/publication. Accessed 21 january 2019

Uyarra MC, Borja Á (2016) Ocean literacy: a "new" socio-ecological concept for a sustainable use of the seas. Mar Pollut Bull 104:1–2. https://doi.org/10.1016/j.marpolbul.2016.02.060

van Leeuwen CJ (2013) City blueprints: baseline assessments of sustainable water management in 11 cities of the future. Water Resour Manag 27:5191–5206. https://doi.org/10.1007/s11269-013-0462-5

West RD (2004) Ocean literacy is key to preserving our oceans and coasts. Mar Technol Soc J 38(4):68–69

Witze A (2017) Ocean ecosystems mapped in unprecedented 3D detail. Nature 541(7635):10–11

Chapter 2
Environmental Education Policies Applied to Marine and Coastal Environments in Brazil: Gaps and Challenges

Alexandre de Gusmão Pedrini

Abstract If compared to other countries around the world, Brazil has developed a rather rich and advanced environmental legislation, including public policies (PPs) related to environmental education (EE) and the coastal and marine zones. Thus, the country can be considered as a model, although there isn't any PP in EE directed exclusively to coastal and marine areas yet, which can be considered as a great challenge to the improvement of EE in the Brazilian coast. In this context, this chapter aims: (a) to briefly present the only two Brazilian federal PPs exclusively oriented to EE; (b) point out 12 Brazilian emblematic federal PPs that can potentially cover the coastal and marine zones; and (c) discuss the main gaps and challenges related to the PPs and the Coastal and Marine Environmental Education (CMEE) in Brazil.

Keywords Coastal and Marine Environmental Education · Public policies · Marine management · Brazilian marine biodiversity

2.1 Introduction

Planet Earth is collapsing. Scientists have been proclaiming the end of life, as we know it today, for the coming decades (IPCC 2014). The democratic formulation, approval, and effective execution of socioenvironmental public policies (PPs) by the governments in each country are urgently needed to face this scenario. Parallel to this, multilateral agreements, articulating several countries with similar or bioregional cultural identities, need to be approved with immediate validity. The nefarious effects of human action on the environment are not limited to the borders of countries since they can migrate from one country to another through rivers, the air, or the ocean. For example, the IPCC (2014) points out that global warming is one of

A. G. Pedrini (✉)
Vegetal Biology Department, Rio de Janeiro State University (UERJ), Maracanã, Rio de Janeiro, Brazil
e-mail: pedrini@uerj.br

N. P. Ghilardi-Lopes, F. A. S. Berchez (eds.), *Coastal and Marine Environmental Education*, Brazilian Marine Biodiversity,
https://doi.org/10.1007/978-3-030-05138-9_2

the best-known scourges of global climate change of anthropic origin that will cause sea level rise and, consequently, will flood fertile tillable land, herd pastures, and cities (Berchez et al. 2008, 2016).

In this context, Giddens (2009) proposes a global policy for coping with the development model of post-industrial society that is contrary to the IPCC recommendation (2014). The current model proposed by the IPCC (op. cit.) privileges the adaptation and accommodation to the current model of capitalist development of the unbridled consumption of industrialized products instead of prevention. Included in this model is the scheduled obsolescence of products that feedback unnecessary consumption of material goods from non-renewable natural resources. Giddens (2009), on the other hand, proposes a global policy model that is not based only on the spontaneous reduction of greenhouse gas emissions such as that of the IPCC (2014), which seems to be not working. The author argues that, in the process of contemporary globalization, the social and environmental issues would only be consolidated through transnational stakeholders. The Brazilian political scientist Moura (1994), almost two decades before, already shared the Guiderian rationale on global policies for environmental issues. This may be the path despite the unsuccessful attempts by the United Nations (UN) in the past.

The quest for transnational and global policies to solve the planet's problems became clear after the end of the Second World War, with the creation of UN. With it, the United Nations Educational, Scientific and Cultural Organization (UNESCO) was created together with an Environmental Education Unit (EEU), which essentially supported events and edited dozens of important publications in its International Environmental Education Program (Pedrini 2011).

In the year 2000, the UN promoted the Millennium Summit, bringing together leaders from 189 countries. The Millennium Declaration, emanated from this meeting, contained eight objectives, one of which was about the environment. Feres and Reis (2014) pointed out that the chapter of the Declaration related to environmental issues (Chap. 7) was poorly presented and could be doomed to failure due to a lack of seriousness. These authors also mentioned that in September 2011 the United Nations created a high-level panel for the formulation of a development Agenda for the period after 2015. Scientists then performed a critical review of the Declaration's objectives, verifying their gaps and challenges. It was observed that, although the objectives deal with social, economic, and environmental issues, these three dimensions were not considered with the same weight and were not well articulated. For instance, the issue of hunger was not properly integrated with sustainable agricultural practices and the rational management of water resources. In addition, there was not a clear articulation between environment and education.

In parallel, UNESCO has established the United Nations Decade of Education for Sustainable Development (2005–2014). The EE approach derived from it (Environmental Education for Sustainable Development—EESD) confronted the Environmental Education for Sustainable Societies and Global Responsibility (EESS) originated from the Treaty on Environmental Education for Sustainable Societies and Global Responsibility (TEESS), published in 1992. Meira and Sato (2005) presented the assumptions of each EE strand and made clear that EESS was the model predominantly adopted in Latin America. EESS is based on: (A) an emerging discourse of organized social movements and networks of social organization (as opposed to international creditors such as the World Bank, International Monetary Fund, and UNESCO); (B) leadership through dialogically open participatory communities (as opposed to companies, decision-makers, and opinion makers); (C) the central problem to be tackled would be social exclusion and negative environmental impacts (as opposed to fight population growth); (D) proposals of public policies, regulated market, and real democracy (as opposed to clean technologies, free market, and formal democracy); and (E) permanent EE as opposed to a 10-year proposal of EESD. Complementarily, in terms of transnational planning, EESS presents a great identity with the Latin-American and Caribbean Program of Environmental Education (Programa Latino-Americano e Caribenho de Educação Ambiental—PLACEA), which was originated in Rio-92 and was edited by the managing body of the National Environmental Education Policy. In this program, protection of coastal and marine areas and the efforts for social equality and poverty reduction were represented as priority actions (Brasil/MMA/MEC 2005). The PLACEA was widely discussed by representatives such as environment ministers at various events held in Venezuela, Cuba, and Panama. Nowadays, PLACEA brings together several countries: Argentina, Brazil, Chile, Costa Rica, Cuba, El Salvador, Guatemala, Honduras, Jamaica, Mexico, Nicaragua, Panama, Peru, the Dominican Republic, and Venezuela.

In general, and especially in the Latin-American context, the research on Public Policies for Environmental Education is poorly developed as a scientific area of expertise. Herrera (1972), for instance, classified PPs as explicit (approved laws in parliaments) or implicit ones (internal regulations of those that implement explicit public policies). This classification may explain contradictions in the application of the PPs, since their operationalization may not actually contemplate the objectives of the law. Two other important classifications are those derived from Layrargues' thinking (2002): (a) authoritarian (when imposed by the state or emanating from the power in force) versus democratic (when coming from lawsuits or public calls or social movements); (b) institutionalized (those with funded programs, plans, projects and actions, concrete results, and periodic assessments) versus non-institutionalized (those that are deprived of continued programs, plans, and actions and with eventual funding, presenting results, depending on who is managing the executing unit). Layrargues (2002) considered the Brazilian National Environmental Education Policy (PNEA), for example, as explicit, authoritarian, and institutionalized. In addition, Dusi (2006) evaluated 43 Brazilian EE programs and identified a community-oriented pattern of conduct, which was only sensitizing, despite their declared objective of an EE, which would be transforming.

It is common sense that Brazil has developed a rather rich and advanced environmental legislation (Drummond and Barros-Platiau 2006), and the country can be considered as a model for other countries around the world, mainly Latin-American ones. In this context, the objectives of the present chapter are: (a) to briefly present the only two Brazilian federal PPs exclusively oriented to EE; (b) point out 12 Brazilian emblematic federal PPs that can potentially cover the coastal and marine zones, according to the author's personal experience; (c) discuss the main gaps and challenges related to the PPs and the Coastal and Marine Environmental Education (CMEE) in Brazil. Both Herrera's (1972) and Layrargues' (2002) typologies will be used in the present chapter, whenever possible, to characterize policies and programs.

2.2 Brazilian Environmental Education (EE)Public Policies (PPs)

Several international EE events have influenced PPs in Brazil since the Stockholm Conference in 1972 and the Belgrade Conference in 1975. Pedrini (2011) considered that three of these events are fundamental and widely known by the name of the cities where they occurred: Tbilisi (1977), Moscow (1987), and Rio de Janeiro (1992). In each one, the conceptual characteristics of EE improved.

Two Brazilian federal PPs exclusively oriented to EE will be presented in this section: National Environmental Education Policy (PNEA) and the Brazilian National Environmental Education Program (ProNEA). Some authors studied PNEA and ProNEA under different perspectives (for more information on those PPs, see Saito 2002; Sorrentino et al. 2005; Barbosa 2008; Loureiro 2009; Nery-Silva and Santana 2016), which will be captured in the text below.

The PNEA (federal law 9.795/1999), established on 1999, is an implicit, authoritarian, and institutionalized PP which does not reflect the aspirations of environmental educators of that time (Brasil 1999). Pedrini (1996, 2004) and Layrargues (2002) drew deep criticism about the precocity and casuistry of its approval by the Brazilian National Congress. PNEA was regulated by Decree 4.281 of June 15, 2002, which created the Management Council, responsible for coordinating the PNEA and headed by the Brazilian Ministries of the Environment and Education. This articulation allowed a perspective of integration between the two instances of EE (education and environment).

ProNEA is an explicit and democratic PP, which is institutionalized in the Brazilian federal sphere. This program was put under public consultation in 2004 and received several improvements by around 800 environmental educators (Brasil 2014). In fact, ProNEA has been the guiding PP for the practice of Brazilian environmental educators. The different versions of ProNEA were based on TEESS, which was approved at the I International Conference on Environmental Education in Rio-92. ProNEA presents different conceptions under different theoretical lines that coexist in the country. The transformative and emancipatory environmental education (TEEE) adopted in ProNEA is that of Paulo Freire, which is characterized as liberating, dialogic, problematizing, transforming, and politically and financially emancipatory. TEEE still needs to be widely developed in Latin-American coastal and marine areas, as well as throughout the Brazilian territory.

In 2017, the IX Brazilian Forum of Environmental Education, held in the city of Balneário de Camboriú in the state of Santa Catarina (Brazil), brought together the environmental educators of the Brazilian Environmental Education Network (REBEA), which encompasses dozens of other territorial and thematic networks in Brazil. In this event, the suggestions for the improvement of the ProNEA (Brasil 2014) made by hundreds of Brazilian educators from almost the whole country were approved. REBEA emphasized coastal and marine ecosystems as exemplified by the conference "EE Policies for the Conservation and Sustainable Use of Coastal and Marine Environments" (Pedrini 2017).

Several territorial networks are linked to REBEA and bring together ordinary people from the public and all those who wish to participate in environmental education, such as teachers, students, academics, and the ones that have seat in several instances of power, as the Rio de Janeiro State Council of the Environment (CONEMA).

CONEMA is an organ of the Environmental Secretariat, which deliberates about PPs and environmental rules. The Rio de Janeiro Environmental Education Network (REARJ), for instance, channels some of the popular pressure that still exists in the country for the creation of PPs and environmental control legislation. On July 26th, 2018, the Rio de Janeiro State Program on Environmental Education, an explicit and democratic PP, was approved by CONEMA.

2.3 Socioenvironmental Public Policies in Brazil

Brazil has several PPs related to environmental issues. However, these policies, in general, do not include EE and, when they do, the focus is almost only related to the aspect of conservation of organisms, ecosystems, or nature. The human beings, as a rule, are seen as predatory and excluded from the environment. It was only after many conflicts, with the expulsion of families of traditional communities who had lived in protected areas before their creation, that PPs started to consider the social dimension, such as the "National Policy for the Sustainable Development of Traditional People and Communities"(Decree n° 6.040, established on February 7th, 2007—Brasil 2007).

In this section, 12 Brazilian emblematic federal PPs, which can potentially affect the coastal and marine zones, will be highlighted, as well as their relation to EE.

2.3.1 The National Environmental Policy (Política Nacional do Meio Ambiente: PNMA) (Federal Law n° 6.938/1981)

In the 1970s, after the publication of the "Silent Spring" by Rachel Carson (1962), as the rules for the installation of industries became more restrictive in the USA, a search for countries such as Brazil, which did not have an environmental legislation yet or whose rules were less restrictive than those of the USA National Environmental Policy Act (NEPA), started. Therefore, Brazilian federal parliamentarians—deputies and senators—encouraged and supported by the environmental movement and by environmental educators, formulated the draft of the National Environmental Policy (PNMA). This federal law was approved, after a lot of political effort, in August 31st 1981. Unfortunately, the environmental laws in Brazil were generally created after serious socioenvironmental accidents that caused human deaths or after intense and continuous popular struggle, associated with the action of few environmental parliamentarians. Therefore, PPs are still created by popular pressure or by government casuism (Pedrini 2011).

Even so, the PNMA is a milestone for Brazilian EE because it is the first national PP to clearly include EE as a legal principle (Brasil 1981). At that time, worldwide, events in EE were occurring (Pedrini 2011) and the Brazilian federal government did not create a specific PP for EE. Because of that, the insertion of EE in a national law even with a limited perspective of training human resources in formal and nonformal education for the protection of the environment can be considered a historical milestone. PNMA was innovative for the time and created environmental licensing for companies with the need to carry out environmental impact studies, which consider EE activities as one of the recurrent mitigation strategies considered. This PP is explicit, democratic, and institutionalized by the federal government and is operationalized by The National Environmental Program (PNMA I and II).

2.3.2 The National Environmental Program (PNMA I and II)

The National Environmental Program (phase I—from 1991 to 1998; and phase II—from 2000 to 2004 and from 2009 to 2014) is an explicit, democratic, and institutionalized by federal government PP which has the objective of contributing to the fortification of the main Brazilian environmental institutions and also strengthen the environmental management capacity at federal, state, and municipal levels (Brasil/MMA 2004). Its creation made possible the subsequent implementation of the National Coastal Management Plan (PNGC), financially feasible, according to Prates et al. (2010a).

PNMA I and II also supported the ORLA Project (Nakano 2006) together with the PNGC to produce socioenvironmental assessments and planning of coastlines.

2.3.3 The National Coastal Management Plan (PNGC)

The PNGC (Brasil 1988) is important for the management of the coastal and marine zone. Although the term "Environmental Education" is not used throughout the document, the plan aims at the establishment of an integrated, decentralized, and participatory management of the socioeconomic activities in the coastal zone, to contribute raising the quality of life of its population, and the protection of its natural, historical, ethnic, and cultural heritages, which is accordance with EE principles.

Xavier (1993) advocated a national law for coastal management to replace PNGC. According to the author, it would be easier to institutionalize the issue in the federal government, which would give more relevance to it in relation to other PPs in the dispute for federal funds and, consequently, facilitate its full compliance.

2.3.4 The Brazilian Agenda 21 Program

The international Agenda 21 was created and approved in Rio-92. As a subsequent step, each country was requested to create its own agenda. On February 26th, 1997 the Commission on Sustainable Development Policies and on the Brazilian Agenda 21 (CPDS) was created, with the goals of formulating, coordinating, and implementing the local version (MMA 1998). The Brazilian Agenda 21 is an explicit, democratic, and institutionalized PP, based on the result of several public meetings all over the country. The Brazilian Federal Senate published a book with almost 600 pages reporting the conclusion of the elaboration works (Brasil 2003). Later, the creation local Agenda 21 was encouraged throughout the country and in schools (MMA 2006a). The concept of sustainable development therein explained was the basis of the pluriannual plan that would be initiated by the government for the period of 2000–2003. Then Agenda 21 became a program and was included in the 2004–2007 pluriannual plan.

The Agenda 21 program is another environmental PP with the goal of creating structures of formulation, evaluation, and decision, which entangled different social actors from different cultures, based on units that could be a school, a community, a river basin, or a municipality (MMA 2006a). It is a planning instrument for the development of sustainable societies, in different geographical localities, conciliating methods of environmental protection, social justice, and economic efficiency.

In chapter 17 of the Global Agenda 21, the protection of ocean and seas, the incentive to integrated coastal management and to the formulation of educational, awareness, and information programs were emphasized. The Brazilian Agenda 21 is the only PP that leaves explicit and demands the necessity of the development of education and capacity building in the context of marine and coastal management. In chapter 36 of the Global Agenda 21, the promotion of teaching, awareness, and capacitation of the population was also encouraged. Teixeira (2006) critically analyzed the Agenda 21 as an explicit Brazilian PP and concluded that its educational proposal leads to socioenvironmental un-sustainability. Based on the analysis of 43 EE Brazilian programs, Dusi (2006) recommended that the transforming EE should be incorporated to the methodologies of the Agenda 21. For the author, the transforming EE should be associated with the processes of democratic strengthening and citizenship building.

2.3.5 The National System of Nature Conservation Units (SNUC) (Federal Law n° 9.985/2000)

The Brazilian System of Protected Areas, called National System of Nature Conservation Units (Sistema Nacional de Unidades de Conservação—SNUC – Brasil 2000) is an explicit, authoritarian, and institutionalized PP that proposes two basic types of protected areas (PAs): those allowing the sustainable use of part of their natural resources and those classified as "no-take" or "of integral protection" (Brasil 2002). EE programs, projects, and actions can be performed in both categories, including coastal and marine protected areas. They are implemented with the incentive of the PA lManagement Plan or the Guidelines for Visiting Protected Areas (MMA 2006b). According to Figueiredo et al. (2008), the guidelines of the implicit policy called "National Strategy for Communication and Environmental Education in Protected Areas" (ENCEA) guides EE actions in the Brazilian PAs. According to Pereira (1999) there were 255 coastal and marine PAs in the country, resulting in hundreds of reports of EE actions in Brazilian PAs (see Pedrini et al. 2010; Prates et al. 2010b; Amaral et al. 2014, and Chap. 7 of the present book). It is very important, then, to make an inventory of the EE promotion derived from PPs in coastal and marine PAs in the country in order to properly identify and characterize the contribution of each of them.

2.3.6 The National Policy on Biodiversity (PNBio) (Federal Decree n° 4.339/2002)

The PNBio (Brasil 2002) can be considered a PP which is explicit, democratic, and institutionalized in the Brazilian Ministry of the Environment. The National Biological Diversity Program (Pronabio) and the National Biodiversity Commission (Conabio) regulate it. Among the 11 Pronabio objectives, we can mention the plan to guide the elaboration and implementation of the PNBio through the promotion of partnerships with civil society for the knowledge and conservation of biological diversity, the sustainable use of its components, and the fair and equitable distribution of benefits derived from its use, in accordance with the principles and guidelines of the Convention on Biological Diversity, the Brazilian Agenda 21, and the National Environmental Policy. In addition, Pronabio formulated the Project for the Conservation and Sustainable Use of Brazilian Biological Diversity (Probio). Probio funded an extensive 18-month study and brought together 177 scientists to select priority areas for the conservation of coastal and marine zone. Six hundered eight areas was determined, 74.2% new ones and 25.8% as PAs or indigenous lands. For the new protected coastal areas, the most recommended actions were inspection (324 recommendations) and EE (286 recommendations). Thus, experts have proposed an efficient coping strategy to reverse the destruction of environmental resources and coastal socioenvironmental conflicts. However, the inventory of marine biodiversity and the knowledge of physical oceanography were considered the priorities for the marine areas, whereas actions in EE were put into ninth place. The federal government has also launched, in the context of Probio, public notices for the promotion of EE. In one of these, Saito (2012) reported on EE for biodiversity conservation, covering all Brazilian biomes and the coastal and marine zone. The author presented playful strategies, focusing on schools and public squares. According to Saito et al. (2014), it was possible to train about 150 teachers from 30 schools from north to south of Brazil, involving the coastal and marine areas of the country and using a critical, liberating, dialogic, and transformative education, as preconized by Paulo Freire's line of reasoning.

2.3.7 Ramsar Convention (Federal Decree n° 1.905/1996)

The biologist Diegues (1987) was one of the first scientists to plead an explicit PP for the conservation and sustainable development of the Brazilian coastal zone. He already mentioned the Ramsar Convention and its importance for the conservation of planetary wet environments. In his work, he briefly presented the entire Brazilian coast and its level of degradation and recommended the implementation of environmental education programs in Brazilian coastal ecosystems. Afterwards, Ramsar Convention was fully incorporated into the legal framework of Brazil in 1996, through the enactment of Decree No.1905/96 (Brasil 1996). Ramsar establishes frameworks for national action and for cooperation among countries to promote the conservation and wise use of wetlands in the world. These actions are based on the recognition by the signatory countries of the Convention of the ecological importance and social, economic, cultural, scientific, and recreational value of such areas and the well-being of the human populations that depend on them.

2.3.8 The National Policy for the Resources of the Sea (PNRM) (Federal Decree n° 5377/2005)

The PNRM was formulated for disciplining the exploitation of natural marine assets. It is explicit, authoritarian, and institutionalized. PNMR was updated in 2005 (Brasil 2005; Prates et al. 2010a) and is intended to guide the development of activities aimed at the effective use and exploitation of the living, mineral, and energy resources of the Territorial Sea, the Exclusive Economic Zone, and the Continental Shelf. Its purpose is in accordance with national interests, in a rational and sustainable way for the socioeconomic development of the country, generating employment and income and contributing to social insertion.

The PNRM aims essentially: (a) to establish principles and objectives for the elaboration of government plans, programs, and actions in the field of human resources training activities; (b) to develop marine research, science, and technology; and (c) to promote the exploitation and sustainable use of marine resources.

One of the consequences of PNRM was the elaboration of the Sectorial Plan for the Resources of the Sea (Plano Setorial para os Recursos do Mar—PSRM) with the objective of knowing and evaluating the potential of the sea, as well as monitoring the resources and the oceanographic phenomena and the climate of the marine areas under national jurisdiction and interest, aiming at the management, sustainable use of these resources, and fair distribution of the benefits arising from such use. The IX PSRM, which runs from 2016 to 2019, encourages the participative management model integrated by Ministries, development agencies, academic and scientific communities, and private initiative, both in its elaboration as in the execution of joint actions and in line with the guidelines of PNRM and other policies and related plans.

There is a law project (Projeto de Lei 6.969/2013) in the federal parliament that proposes to institute the National Policy for the Conservation and Sustainable Use of the Brazilian Marine Biome (PNCMar – Brasil 2013), which proposes the incentive to environmental education and awareness of the population on the importance of conservation, recovery, and management of coastal and marine resources.

2.3.9 The Maritime Mentality Program (PROMAR)

PROMAR is organized by the Interministerial Commission for the Resources of the Sea (CIRM – Brasil 2018) and aims: (1) to propose the inclusion of themes related to the sea in the curricula of primary and secondary education; (2) to increase the organization of itinerant exhibitions and lectures on marine subjects; (3) to expand the distribution of material for the dissemination of CIRM activities; (4) to strengthen the bonds between the collectivity and the sea, through the dynamization of the oceanographic museums; and (5) extend to 2.4 million the estimated number of people reached annually by actions related to the development of maritime mentality.

It is the only Brazilian PP exclusively devoted to ocean literacy (see more on this subject on Chap. 1).

2.3.10 The National Policy for the Sustainable Development of Traditional People and Communities (PNPCT) (Federal Decree n° 6.040/2007)

The PNPCT seeks to promote the sustainable development of Traditional People and Communities, with an emphasis on the recognition, strengthening, and guarantee of their territorial, social, environmental, economic, and cultural rights, with respect and appreciation for their identity, their forms of organization, and their institutions. It is a Brazilian PP that is effectively concerned with sociodiversity. It is an action of the federal government that was not emanated from the National Congress but rather pleaded by social movements. In this way, this PP can be considered explicit, democratic, and institutionalized in the federal government. One of the specific objectives of this policy is to value and guarantee the traditional forms of education and strengthen dialogic processes, so that those who live in the coastal and marine areas can have their culture preserved (Brasil 2007). According to Diegues (1999), it is very important to preserve the culture of traditional populations, since they know techniques for the cultivation of species for their survival, which are fully integrated into the native coastal Brazilian biodiversity. They also hold expertise on several aspects of oceanographic processes, from currents to areas with high fisheries production.

2.3.11 Conservation Program for Brazilian Coral Reefs

According to Prates et al. (2010a), Brazil is the only country in the South Atlantic that has coral reefs, and the major reef builders are species endemic to Brazilian waters (Leão et al. 2016), occurring along approximately 3000 km of coastline. One conservation action related to Brazilian coral reefs is the Campaign for Conscious Conduct in Reef Environments, which is implicit, democratic, and institutionalized in the Ministry of the Environment. Prates et al. (2010b) made a rich account of this EE campaign, explaining all its methodology and the interesting results obtained with the 8 h course with campaign material (booklet, poster, sticker, and DVD). This program allows for the integration with projects involving traditional populations and caiçaras (traditional fishermen), such as marine ecotourism (Rhormens et al. 2017; Pedrini et al. 2016a, b; Pedrini 2017, 2018).

2.3.12 Public Policies in Science, Technology, and Innovation (S, T, and I) in Brazil

Brazilian PPs in S, T, and I also support EE projects and actions through scientific communication programs or science and technology funds. There are several policies, programs, and plans, but two examples of funding will be presented in this section. One was a project of the Probio, which was funded also by a call for the popularization of science. This project reported on EE for biodiversity conservation, covering all Brazilian biomes and the coastal and marine zone (Saito 2012). The other is the financial support for a network of scientists involving several institutions and funded to rationalize financial costs. This funding through a PP in S, T, and I for EE was directed to the Brazilian Monitoring Network for Coastal Benthic

Habitats (ReBentos) from 2011 to 2015. The Brazilian National Council for Scientific and Technological Development (CNPq), the Research Program on Biodiversity Characterization, Conservation, Restoration and Sustainable Use of the São Paulo Research Foundation (BIOTA-FAPESP), the Coordination for the Improvement of Higher Education Personnel (CAPES), and the Brazilian Innovation Agency (FINEP) funded ReBentos. The main objective of the network is the integrated implementation of long-term monitoring studies of the benthic habitats of the Brazilian coast considering global climate change (GCC). ReBentos brings together 166 researchers from 57 national and international teaching and research institutions acting in the 17 Brazilian coastal states (Turra and Denadai 2016). It is organized into working groups (WG): (A) rhodolith beds; (B) rocky shores; (C) environmental education; (D) estuaries; (E) vegetated submerged funds; (F) mangroves and marshes; (G) beaches; and (H) coral reefs. It is possible to notice that environmental education WG is empowered, under equal conditions as the other WGs and not situated within them. Nonetheless, EE WG can act transversally to all of them and propose independent projects, although the WG follows its path thematically linked to GCC and its effects in coastal and marine regions (Berchez et al. 2016).

2.4 Gaps and Challenges

As exposed, Brazil has many explicit or implicit, democratic or authoritarian, institutionalized, or not institutionalized socioenvironmental public policies, directly and indirectly covering the coastal and marine zones, but in general, they do not emphasize the need and role of EE. In addition to the socioenvironmental PPs, there are many others, which aim the environmental conservation and simultaneously affect the coastal and marine areas, but they were not dealt with in the present chapter. However, through the analysis of the PPs listed here and in the literature cited it was possible to identify gaps and challenges to foster CMEE in Brazil (Table 2.1).

One very important gap, related to the active participation of the society in the PPs, is the lack of guarantee of representativeness in the instances of power. In many cases, the population has seats in important Councils, such as those of the marine protected areas, but since there is generally no parity between society and government, the population's litigation usually loses in the collective voting. PPs on coastal and marine zones, whether authoritarian or not, implicit and not yet institutionalized at different organizational levels of government, should be presented to the population for consideration and evaluation. They may be deactivated or enforced in accordance with the lawsuits of the social actors involved or affected and their results need to be given wide publicity.

Another gap evidenced has been the lack of use of the centennial knowledge of traditional populations, supposing that they would not know how to make their survival compatible with the native Brazilian coastal biodiversity.

Table 2.1 Ten main gaps and associated challenges on environmental PPs applied to the coastal zone and marine environments in Brazil

	Gap	Challenges
1	Popular ignorance about the fundamental role of a PP, as well as its life cycle and how to formulate a draft to be submitted to one of the parliaments in the country	Promote short- and medium-term courses for different levels of the population on PPs to enable the population to plead their PP demands to the parliaments of their choice
2	Insufficient number of professionals able to guide the population to formulate PP projects, as well as to teach courses on the subject	Encourage the development of research and teaching groups in explicit and implicit public policies in government universities
3	The absence of a direct channel between the population and the government to know and debate explicit PP projects derived from the government and to debate the PPs originated by the community	Ensure the population that will be affected by a PP that the holding of public hearings is frequent on weekends, in places of easy access and that are widely communicated well in advance
4	Lack of representation, voice, and vote of the traditional populations and racial minorities, in the management of PPs, making it impossible to make available their centennial knowledge in the management of the territory and the natural assets of the coast	Include traditional populations and racial minorities in the formulation, litigation, and management of explicit PPs that have an effect on their territory
5	Lack of knowledge if the EE is (and if it is, how it is) institutionalized in the state and municipal departments of education and the environment on the Brazilian coast	Institutionalize the EE in the state and municipal departments of education and of the environment in the Brazilian coast
6	Lack of information to formulate a macrodiagnosis of socioenvironmental conditions with up-to-date sectoral data from the coastal and marine zones	Operationalize PPs through programs, plans, and sectoral projects along the entire coast
7	Lack of knowledge about the legislation that needs constant updating or its regulation to have the force of law	Regulate all articles of approved legislation on PPs in EE that require this provision to be put into practice
8	Lack of evaluation indicators to evaluate the performance of explicit PPs that are publicly available	Develop methods for evaluating the performance of explicit public policies (PPs)
9	Decentralization and sharing of responsibilities in the distribution of financial resources to the implementation of PPs	Allocate financial resources to the programs and projects of the PPs approved and considered as a priority by the population through the councils in which it is represented and is the majority
10	Lack of knowledge about the total number of EA actions and projects developed in PPs that receive the coastal or marine name or that somehow deal with the theme	Promote a census of federal, state, and municipal governments on how EE actions/projects/programs are implemented in PPs under their jurisdiction that address coastal and marine issues

In addition, the PPs that are already specific and explicit for the coastal and marine zones, even those with biological concerns, should be imbricated with those that are pedagogical in their essence, to allow for the search of identities for the formulation of EE work plans. The mere view of conservation of organisms or coastal and marine ecosystems with human exclusion should be reviewed. Human beings must be reinserted not only as a predator or harvester but also as a vulnerable species.

Explicit public policies concerning both the environment and the environmental education must be evaluated periodically, as stated in their own content. Users of explicit, democratic, and institutionalized PPs as the population, in general, need more effectiveness in the services provided. On the other hand, professors, pedagogical managers, students, and scientists need greater continuity in the promotion and the funding of projects of EE which affect the coastal and marine zones.

The gaps identified in this chapter and many others evidenced in the literature about PPs about coastal and marine zone or about EE are difficult to solve considering the continental dimensions of a country as Brazil. Besides, the ideal teams of environmental educators are multidisciplinary and, thus, expensive to maintain.

Transportation from south to north in the country through airfare with daily rates makes projects more expensive, since EE projects must correspond to a permanent process in temporal terms, being territorially wide and cross-disciplinary.

The challenges to extinguish the gaps identified here are enormous and their achievement will be possible only at the medium and long term, if properly funded. These challenges demand a proper organizational structure with qualified personnel, since working with the coastal and marine zones requires specific knowledge. The professionals that the country already has with graduated and trained personnel are biologists, physicists, chemists, and oceanographers. The most pre-eminent challenges for PPs could be faced with organization, physical structure, and qualified personnel.

Added to this, there is a lack of knowledge about the PPs that demand the training of citizens in short-, medium-, and long-term courses at different school levels about PPs so that they can plead their demands to the parliaments of their choice. If the citizens can participate in the formulation of PPs and get to know them in detail, it is easier for them to follow up and collect their compliance from the government that executes the legislation. It is only with such attitudes that we will actually have a true democracy in the country with the direct participation of the society as a social control actor. This new approach would also contribute to preventing corruption.

The population's lack of knowledge about how to enforce explicit PPs in EE and in the coastal and marine zones occurs frequently in the country. There is also a need for a broad public training program so that ordinary citizens can formulate PP projects on topics that affect their daily lives. Associated with this, there is also the lack of a large number of public agents trained to provide this qualification throughout the country.

The challenge of facing the lack of expertise in the field of PP in coastal and marine EE in the country could be facilitated with financial investment in public universities in all states of the federation. A public notice for creation of a higher degree in public policy could be created, compulsorily associated with graduate courses at the MSc and PhD level with research lines in explicit and implicit democratic PPs for institutionalization in public bodies.

The challenge of including traditional populations, quilombolas (descendants from African slaves), caiçaras, anglers, and population and racial minorities in the formulation, litigation, and management of explicit PPs that influence their territory is one of the biggest to be faced, since they do not form a unit and each of these collectives has a leadership. Therefore, it will be a huge challenge to create an understanding to establish communication channels with all leaders in order to talk on an equal level with everyone.

The difficulty of communication between the coastal zone citizen and the municipal and state governments mainly affects the charge of the fulfillment of the PPs. One of the channels provided in Brazilian PPs is the holding of public hearings. However, they need to be widely communicated, well in advance, and occur preferably on Sundays and national holidays and in places of easy access to all. The public universities could highlight teachers to assist the society, and these professors should receive recognition for this role in their career promotions.

The institutionalization of CMEE in the federal, state, and municipal departments of education and the environment in the Brazilian coast should be automatic considering its thematic identity and the existence of environmental programs in government sectors such as the Public Administration Environmental Agenda (A3P—MMA 2018).

It is necessary to promote a census of the actions/projects/programs of EE in the PPs of the coastal zone and marine by the federal, state, and municipal governments. By their dimensioning, there will be conditions to make a contemporary national macrodiagnosis, which will make possible short-, medium-, and long-term government decisions for the national education and environment related to the Brazilian coast and marine environments.

In legislation, many articles can be considered self-explanatory when the PP is published. However, many of them require regulation by another legal act, which can take years to be published or will never be approved, making the initial legislation for which that PP was created unfeasible. Thus, the population needs a lawyer regularly registered with the Brazilian Order of Attorneys to advise them and to know if the PP they helped to formulate or that affects them directly is being fulfilled or not.

One of the ever-present articles in laws approving PPs, including those in EE, is the need to develop and carry out an evaluation of their effectiveness. However, even the evaluations of explicit PPs in EE rarely are widely disseminated to the population affected by them, and are only disseminated in scientific journals, thematic meetings, or low-run books.

Another challenge to be faced is to be aware that in all Councils in which society is represented there must be parity with other represented entities so that the population has the possibility to see their lawsuits approved when proposed and voted. Thus, when proposing public or private funds, the society must be able to receive them.

The challenge of the operationalization of PPs derived from parliamentarians or government through programs, plans, and sectorial projects along the entire coast, allocating budget funds from the multi-year plan is one of the most herculean of all. Disputes over public funds are extremely fierce, and certain politicians with greater bargaining power in parliament often only make funds available to operationalize PPs through parliamentary amendments.

The almost continental size of Brazil has always discouraged national censuses in the face of their high financial costs and the complexity of being reached all over the country. Only those that are required by law as the recapturing of the population are periodically made. Even so, some partial surveys are made, based on respondents' goodwill. One of them was the work of Veiga et al. (2005) on environmental education carried out in primary schools. The authors got the answer from a few secretariats and yet they were able to determine that 65,000 schools in the country were conducting EE projects in 2004. A census of the number of EE projects in the Brazilian coastal and marine area still needs to be carried out.

Although it is possible to identify gaps and challenges related to Brazilian PPs, socioenvironmental issues approaches based only on the technical perspective, without the transformation and the political and financial emancipation of humankind by education, will result in failure (Quintas 2004). A critical, liberating, dialogic, and transformative education allows, even, planning and building a new socialist planet (Freire 1996). This new worldview will present social and environmental justice, collective empowerment, alterity, equal income distribution, problematizing dialogues, love, sustainable use of social and environmental goods, and intelligent use of social and environmental services (Freire 1996). Berchez et al. (2016), Pedrini et al. (2014, 2016a, b), Rhormens et al. (2017), and Pedrini et al. (2018 – this book, Chap. 6) argue that only a transformative and emancipatory environmental education will allow the combination of progress of the technique with the humanistic development of the human beings, in the coastal and marine zones as in other regions of the planet.

The formulation, discussion, and effective approval of national, regional, and local public policies are also up to Brazil, as well as the negotiation of transnational and global PPs, which should represent with legitimacy the yearnings of socioenvironmental agendas alleged by the communities of their coastal and marine territories.

It will be possible to reverse the planetary socioenvironmental destruction with democratic, explicit, and institutionalized public policies dealing with transformative and emancipatory environmental education in the coastal and marine zones, both in Brazilian territory and in other countries.

Acknowledgements To my dear friend Jalton Gil Torres Pinho for the English version (part) of this work and to Maria Emília Cavalcanti and Laercio Vinhas for the Professional opportunities granted. To my mother Nair Batista de Gusmão (In memoriam) and my dear wife Rosana Nunes Pedrini. For the friendship of Flávio Berchez and Daniel Shimada Brotto. To Prof. Dr. Alexander Turra (USP) and Profa. Dra. Natalia Ghilardi-Lopes (UFABC) for the review of the chapter.

References

Amaral FMD, Santos MFAV, Melo KV et al (2014) The role of environmental education in changing school students' perceptions of and attitudes toward coral reefs in the Fernando de Noronha Archipelago, Brazil. Rev Gest Costeira Integr 14(4):1–8

Barbosa LC (2008) Políticas públicas de educação ambiental numa Sociedade de risco: tendências e desafios no Brasil. In: IV Encontro Nacional da Anppas, 4, 5 e 6 de junho de 2008, Brasília (DF), Brasil

Berchez FAS, Oliveira Filho EC, Amâncio CE, Ghilardi NP (2008) Possíveis impactos das Mudanças Climáticas Globais nas comunidades de organismos marinhos bentônicos da costa brasileira. In: Buckeridge MS (Org.) Biologia e Mudanças Climáticas no Brasil. RiMa, São Carlos, pp 167–180

Berchez FAS, Ghilardi-Lopes NP, Correia MD et al (2016) Marine and coastal environmental education in the context of global climate changes - synthesis and subsidies for ReBentos (Coastal Benthic Habitats Monitoring Network). Braz J Oceanogr 64(spe2):137–156. http://doi.org/10.1590/S1679-87592016093 2064sp2

Brasil (1981) Lei nº 6.938, de 31 de agosto de 1981. Dispõe sobre a Política Nacional do Meio Ambiente, seus fins e mecanismos de formulação e aplicação, e dá outras providências. Diário Oficial da União, Brasília

Brasil (1988) Lei nº 7.661, de 16 de maio 1988. Institui o Plano Nacional de Gerenciamento Costeiro e dá outras providências. Diário Oficial da União, Brasília

Brasil (1996) Decreto n° 1.905, de 16 de maio de 1996. Promulga a Convenção sobre Zonas Úmidas de Importância Internacional, especialmente como Habitat de Aves Aquáticas, conhecida como Convenção de Ramsar, de 02 de fevereiro de 1971. Diário Oficial da União, Brasília

Brasil (1999) Lei nº 9.795, de 27 de abril de 1999. Dispõe sobre a educação ambiental, institui a política de educação ambiental e dá outras providências. Diário Oficial da União, Brasília

Brasil (2000) Lei nº 9.985, de 18 de julho de 2000. Regulamenta o art. 225, § 1o, incisos I, II, III e VII da Constituição Federal, institui o Sistema Nacional de Unidades de Conservação da Natureza e dá outras providências. Diário Oficial da União, Brasília

Brasil (2002) Decreto nº 4.339, de 22 de Agosto de 2002. Institui princípios e diretrizes para a implementação da Política Nacional da Biodiversidade. Diário Oficial da União, Brasília

Brasil (2003) Agenda 21: Conferência das Nações Unidas sobre Meio Ambiente e Desenvolvimento. Ministério do Meio Ambiente, Brasília

Brasil/MMA (2004) Programa Nacional do Meio Ambiente – PNMA II. Ministério do Meio Ambiente, Brasília

Brasil/MMA-MEC (2005) Programa Latino-americano e Caribenho de Educação Ambiental. Série Documentos Técnicos, 5, Brasília

Brasil (2005) Decreto nº 5.377, de 23 de Fevereiro de 2005. Aprova a Política Nacional dos Recursos do Mar - PNRM. Diário Oficial da União, Brasília

Brasil (2007) Decreto n° 6.040, de 7 de fevereiro de 2007. Institui a Política Nacional de Desenvolvimento Sustentável dos Povos e Comunidades Tradicionais. Diário Oficial da União, Brasília

Brasil (2013) Projeto de Lei n° 6.969 de 2013. Institui a Política Nacional para a Conservação e o Uso Sustentável do Bioma Marinho Brasileiro (PNCMar) e dá outras providências. Brasília

Brasil (2014) Programa Nacional de Educação Ambiental (ProNEA). 4 ed. Ministério do Meio Ambiente, Brasília

Brasil (2018) Programa de Mentalidade Marítima da Comissão Interministerial para os Recursos do Mar (CIRM). Available at: https://www.marinha.mil.br/secirm/promar. Accessed 23 June 2018

Carson R (1962) Silent Spring. Houghton Mifflin, Boston

Diegues AC (1987) Conservação e desenvolvimento sustentado de Ecossistemas litorâneos no Brasil. In: Simpósio de Ecossistemas do Sul e Sudeste do Brasil, 11 a 16 de abril de 1987, pp 196–245

Diegues ACS (1999) Human populations and coastal wetlands: conservation and management in Brazil. Ocean Coast Manag 42:187–210. https://doi.org/10.1016/S0964-5691(98)00053-2

Drummond J, Barros-Platiau AF (2006) Brazilian environmental Laws and Policies, 1934–2002: a critical overview. Law Policy 28(1):83–108. https://doi.org/10.1111/j.1467-9930.2005.00218.x

Dusi RLM (2006) Estudo sobre programas de Educação Ambiental no Brasil e a questão da sustentabilidade ambiental. Tese de Doutorado em Ecologia, Universidade de Brasília

Feres J, Reis E (2014) A agenda de desenvolvimento pós-2015: a questão ambiental no contexto do desenvolvimento sustentável. In: Monasterio LM, Neri MC, Soares SSD (eds) Brasil em Desenvolvimento 2014; estado, planejamento e políticas públicas, vol 2. Ipea, Brasília, pp 17–36

Figueiredo ALCB, Lima A, Mendonça FC et al (2008) Diretrizes para Estratégia Nacional de Comunicação e Educação Ambiental em Unidades de Conservação – ENCEA. Ministério do Meio Ambiente, Brasília

Freire P (1996) Pedagogia da autonomia: saberes necessários à prática educativa. Paz e Terra, São Paulo

Giddens A (2009) The politics of climate change. Polity Press Ltd, Cambridge

Herrera A (1972) Social determinants of science policy in Latin America; explicit science policy and implicit science policy. J Devel Stud 9(1):19–37. https://doi.org/10.1080/00220387208421429

IPCC (2014) Cambio climático 2014: impactos, adaptación e vulnerabilidade, resumen para responsables de políticas. Organización Meteorológica Mundial, Genebra. Available at: https://www.ipcc.ch/pdf/assessment-report/ar5/wg2/ar5_wgII_spm_es.pdf. Accessed 27 February 2015

Layrargues PP (2002) A conjuntura da institucionalização da Política Nacional de Educação Ambiental. OLAM Ciênc Tecnol 2(1):1–15

Loureiro CFB (2009) Mundialização do capital, sustentabilidade Democrática e políticas públicas: Problematizando os caminhos da educação ambiental. Ambient Educ 14(1):11–22

Leão ZMAN, Kikuchi RKP, Ferreira BP et al (2016) Brazilian coral reefs in a period of global change: a synthesis. Braz J Oceanogr 64(spe2):97–116. http://doi.org/10.1590/S1679-875920160916064sp2

Meira P, Sato M (2005) Só os peixes mortos não conseguem nadar contra a correnteza. Rev Educ Pública 14(25):17–31

Ministério do Meio Ambiente (1998) Agenda 21 - o caso do Brasil: perguntas e respostas. Ministério do Meio Ambiente, Brasília

Ministério do Meio Ambiente (2006a) Passo a passo da agenda 21 local. Ministério do Meio Ambiente, Secretaria de Políticas para o Desenvolvimento Sustentável, Brasília

Ministério do Meio Ambiente (2006b) Guia para visitação de Areas Protegidas. Ministério do Meio Ambiente, Brasilia

Ministério do Meio Ambiente (2018) O que é A3P? Available at: http://www.mma.gov.br/responsabilidade-socioambiental/a3p. Accessed 19 September 2018

Moura AS (1994) Estado, transnacionalidade e políticas globais. Rev Serv Público 59(2):245–252. https://doi.org/10.21874/rsp.v59i2.148

Nakano K (2006) Projeto Orla. Instituto Pólis, São Paulo

Nery-Silva AC, Santana LC (2016) Políticas públicas e educação ambiental: concepções e contextos nas pesquisas acadêmicas brasileiras. Rev Eletrônica Mestr Educ Ambient 33(2):158–176. https://doi.org/10.14295/remea.v33i2.5654

Pedrini AG (1996) Uma crítica à proposta legislativa para a criação da política brasileira de educação ambiental. In: Anais do I Seminário Educação Ambiental e a Nova Ordem Mundial, Casa da Ciência, Universidade Federal do Rio de Janeiro, 29 a 31 de outubro de 1996, pp 110–112

Pedrini AG (2004) The National Public Policies with Environmental Education in Brazil: evolutions and perspectives. In: Azeiteiro U, Gonçalves F, Leal W, Morgado F, Pereira M (Org.) World Trends in Environmental Education. Peter Lang, Frankfurt, pp 115–122

Pedrini AG (2011) Trajetórias da Educação Ambiental. In: Pedrini AG (Org.) Educação Ambiental; reflexões e práticas contemporâneas. Vozes, Petrópolis, pp 25–89

Pedrini AG (2017) Ecoturismo marinho sustentável como educação ambiental transformadora e emancipatória em unidades de conservação marinha no Brasil: uma proposta inovadora da universidade à sociedade. In: Guerra AFS, Figueiredo ML (Org.) Diálogo de Saberes e Fazeres; uma releitura dos 25 anos da trajetória da educação ambiental Brasileira. ICEP Editora, São José, pp 298–302

Pedrini AG (2018) Marine ecotourism in Brazil: from origins to implementation feasibility. Rev Bras Ecot 11(2):506–540

Pedrini AG, Brotto DS, Pimentel DS, Behrends E, Moraes AJ (2016a) Transformative and emancipatory environmental education by marine ecotourism in the marine environmental protection area of Armação dos Búzios, Rio de Janeiro, Brazil. Rev Bras Ecot 9(3):445–470

Pedrini AG, Rhormens MS, Brotto D (2016b) Educação Ambiental Emancipatória pelo Ecoturismo Marinho de Base Comunitária; uma Proposta Metodológica com Sustentabilidade Socioambiental. In: Araujo MI, Santana CG, Nepomuceno AOL (Org.) Educadores Ambientais: caminhos para a práxis. Ed. Criação, Aracajú, pp 47–64

Pedrini AG, Maneschy FSA, Silva VG, Campos PH, Costa C, Andrade-Costa E, Newton T (2010) Projeto EduMar – Educação/Interpretação Ambiental Marinha para o mergulho recreativo. In: Pedrini AG (Org.) Educação Ambiental Marinha e Costeira no Brasil. Eduerj, Rio de Janeiro, pp 143–165

Pedrini AG, Ursi S, Berchez F, Correia MD, Sovierzoski HH, Mochel F (2014) Metodologias em Educação Ambiental para a conservação socioambiental dos ecossistemas marinhos. In: Pedrini AG, Saito CH (Org.) Paradigmas Metodológicos em Educação Ambiental. Vozes, Petrópolis, pp 132–151

Pereira PM (1999) Sistematização das Informações relativas às Unidades de Conservação das Zonas Costeira e Marinha do Brasil. Ministério do Meio Ambiente, Brasília

Prates AP, Gonçalves MA, Rosa MR (2010a) Panorama da conservação dos ecossistemas costeiros e marinhos no Brasil. Ministério do Meio Ambiente, Brasília

Prates AP, Pereira PM, Duarte AEM, Hazin MC, Ferreira BP, Reinhardt M (2010b) Campanha de conduta consciente em ambientes recifais. In: Pedrini AG (Org.) Educação Ambiental Marinha e Costeira no Brasil. Eduerj, Rio de Janeiro, pp 115–131

Quintas JS (2004) Educação no processo de gestão ambiental: uma proposta de educação ambiental transformadora e emancipatória. In: Layrargues PP (Coord.) Identidades da Educação Ambiental Brasileira. Ministério do Meio Ambiente, Brasília, pp 113–140

Rhormens MS, Pedrini AG, Ghilardi-Lopes N (2017) Implementation feasibility of a marine ecotourism product on the reef environments of the marine protected areas of Tinharé and Boipeba Islands (Cairu, Bahia, Brazil). Ocean Coast Manag 139:1–11. https://doi.org/10.1016/j.ocecoaman.2017.01.022

Saito CH (2002) Política Nacional de Educação Ambiental e Construção da Cidadania: Desafios Contemporâneos. In: Ruscheinsky A (Org.) Educação Ambiental: abordagens múltiplas. Artmed, Porto Alegre, pp 47–60

Saito CH (2012) Popularizando o Probio-Educação Ambiental na Praça e na Escola. Rev Bras Educ Ambient 7(2):83–95

Saito CH, Figueiredo J, Vargas IA (2014) Educação Ambiental Freiriana no contexto de formação de educadores ambientais. In: Pedrini AG, Saito CH (Org.) Paradigmas Metodológicos em Educação Ambiental. Vozes, Petrópolis, pp 71–81

Sorrentino M, Trajber R, Mendonça P, Ferraro Junior LA (2005) Educação Ambiental como política pública. Educ Pesqui 31(2):285–299. http://doi.org/10.1590/S1517-97022005000200010

Teixeira C (2006) Educação e desenvolvimento sustentável na Agenda 21 brasileira. Ambient Educ 11(1):135–152. http://doi.org/10.5216/ia.v33i1.4237

Turra A, Denadai MR (2016) Linking biodiversity and global environmental changes in Brazilian coastal habitats. Braz J Oceanogr 64(spe2):3–4. http://doi.org/10.1590/S1679-87592016064sp2ed

Veiga A, Amorim ME, Bianco, M (2005) Um retrato da presença da EA no Ensino fundamental Brasileiro: o percurso de um processo acelerado de expansão. Instituto Nacional de Estudos e Pesquisas Educacionais Anísio Teixeira/Ministério da Educação. Textos para Discussão, vol 21, 23p

Xavier R (1993) A Política Nacional de Gerenciamento Costeiro. In: III Simpósio de Ecossistemas da Costa Brasileira, Academia de Ciências do Estado de São Paulo, pp 199–222

Chapter 3
Integration of Marine and Coastal Geo-Biodiversity in Environmental Education Through Earth System Science and Social Learning Approaches

Denise de La Corte Bacci, Maria da Glória Motta Garcia, Christine Laure Marie Bourotte, and Vânia Maria Nunes dos Santos

Abstract In this chapter, we will review the concept of geodiversity, highlighting the geologic characteristics of the rocky shores in Brazil, the main coastal dynamic processes, and the link between geodiversity and biodiversity in the landscape features considering biophysical aspects of the coast. Secondly, we will discuss the role of environmental education in terms of ecological degradation and conflicts in coastal and other environments, such as urban sprawl, segregation, traffic congestion, habitat degradation, urban heat island effect, and water pollution. In this complex system, we understand that education based on earth system science and the social learning approach can help to clarify environmental issues and contribute to citizens' education towards a sustainable society.

Keywords Geodiversity · Environmental education · Social learning · Earth system · Complexity

3.1 Introduction

The interactions between bio- and geodiversity have existed since the emergence of life, about 3.8 billion years ago in geologic time. Throughout its long existence, Earth has been changing continuously. The natural dynamics in time contributed to the constitution of Earth as we know it today. Historical geology aims to understand many processes that operate beneath and upon the Earth's surface as well as the

D. L. C. Bacci (✉) · M. G. M. Garcia · C. L. M. Bourotte · V. M. N. Santos
Núcleo de Apoio à Pesquisa – Geohereditas - Instituto de Geociências, University of São Paulo (USP), São Paulo, São Paulo, Brazil
e-mail: bacci@usp.br; mgmgarcia@usp.br; chrisbourotte@usp.br; vanianunes@ige.unicamp.br

N. P. Ghilardi-Lopes, F. A. S. Berchez (eds.), *Coastal and Marine Environmental Education*, Brazilian Marine Biodiversity,
https://doi.org/10.1007/978-3-030-05138-9_3

most diverse biophysical–chemical interactions in its systems. The overall behavior of the Earth system is strongly shaped by the interactions among its various component systems, such as the atmosphere, hydrosphere, lithosphere, and the inner Earth, but also by life and human activity. These parts are not isolated, each one being related in some way to the others and producing a complex and continuously interacting system. In this context, it is possible to understand all aspects of the succession of life over the vast expanse of geologic time. Species have existed and others have become extinct, leaving fossils as important records of time past. Fossils allow researchers to understand past environmental conditions, giving a more comprehensive view of geologic history and establishing association between present and past species.

The Earth dynamics is recorded in the most diverse features of the landscape, e.g., mountain belts, sedimentary basins, and flat stable areas, as well as different types of minerals, rocks, and ore deposits, constituting what is called as geodiversity. Such records allow us to study, analyze, understand, formulate hypotheses, and produce knowledge about our planet. Understanding the basic geological principles along the way is important to establish relationships between people and nature and comprises a fertile field to develop environmental education initiatives.

Resources are another important value to society because they form the very foundation of modern civilizations. Human presence on Earth is very recent and has triggered over the last two centuries a growing demand for resources and profound impacts on geo- and biodiversity, often irreversible. Humans can dramatically influence geologic processes, which led to the proposition of a new geological age, the Anthropocene (Crutzen and Stoermer 2000—see also Chap. 1). Among many challenges society must confront nowadays, the most worrying are urban sprawl, segregation, traffic congestion, as well as habitat degradation, climate change, and water pollution.

Considering the current environmental complexity in densely urbanized cities on the shoreline, emergent conflicts are arising. Consequently, they are generating dilemmas, inequities, and social inequalities. How should we deal with this social conflict scenario? How can we promote sustainable development, social justice, and public policies?

In this context, in this chapter we deal with examples of natural interactions between geo- and biodiversity and experiences of social learning. Despite its importance, environmental legislation has historically privileged the biotic portion of nature to the detriment of the abiotic portion, especially in the conservation of biodiversity. Since 2008, however, IUCN (International Union for the Conservation of Nature) has been inserting topics related to geodiversity in its agenda, through the inclusion of institutions (Geological Society of Spain, ProGEO), formation of the Group of Specialists in Geological Heritage (2014), inclusion of a chapter on Geodiversity in the Management Manual of Protected Areas (2015) and resolutions (4040/2008, 048/2012, 091/2016, among others). These actions bring, globally, optimistic perspectives regarding the future of the conservation of the physical environment as a base for sustainable management.

Rocky shores in the Southeast of Brazil are exemplified here as places where interaction with geo- and biodiversity occurs and also because is being increasingly affected by urbanization. It covers characteristics of a learning environment that can engage teacher, students, and community in a meaningful discourse about their impact on the planet, and the role place plays in their own well-being (Semken et al. 2017).

We also present in this chapter educational proposals to promote the conservation of bio- and geodiversity and environmental sustainability, which consider all systems' interactions, and can also be applied in the rocky shore environment.

3.2 Geodiversity: The Concept

The term geodiversity emerged in the late 1990s in the context of the territorial management of protected areas, as a complement to the concept of biodiversity, which deals with the biological elements of the natural environment. Since then, the term has been defined in several ways, from the concept of geological variety to a broader definition, which includes the diversity of abiotic nature (Serrano Cañadas and Ruiz Flaño 2007). In a more comprehensive context, geodiversity can be defined as *"the natural variety (diversity) of geological (rocks, minerals, fossils), geomorphological (forms of relief, topography, physical processes), pedological and hydrological features, including systems and landscapes"* (Gray 2013).

The study of geodiversity can consider different scales, ranging from global, which includes continents and oceans, to elemental, which includes atoms and ions. Global scale deals with the history of Earth and plate tectonics, which is fundamental to understand the current geodiversity of our planet. Local scale relates to the materials, processes, and landforms that make up the abiotic portion of the planet and constitute the elements of geodiversity (Fig. 3.1).

The most important and relevant part of geodiversity is the geological heritage, which is composed of places with specific value for human society, including scientific, educational, or cultural value, among others. Rocky shores represent this kind of complex environment and will be described below by their origin and importance related to geo- and biodiversity, as well as social, cultural, and education process.

3.3 The Geodiversity of the Rocky Shores of Brazil

The origin of the South American continental margin is related to the opening of the South Atlantic Ocean, as a result of the breakup of the Western Gondwana Supercontinent, which began in the Jurassic/Cretaceous (about 180 million years ago). Therefore, the lithologic types that make up the Brazilian rocky shores are either associated with this older supercontinent or its rifting.

Fig. 3.1 Geodiversity elements. Terrestrial materials: (**a**) Bicolor tourmaline in pegmatite from Santa Maria do Suaçuí, Minas Gerais. Collection Luiz Paixão, Museum of Geosciences of USP. Photo by: Míriam Azevedo; (**b**) Itabirite, metamorphic rock constituted by quartz- and hematite-rich layers and used as iron ore. Quadrilátero Ferrífero, Minas Gerais; geological processes: (**c**) Coastal-fluvial depositional processes forming estuarine sandbar. Notice the Pleistocene marine terraces in the background. Itaguaré Beach, Bertioga; (**d**) Itú Suite granite showing polished surface and striations caused by passage of glacier in glacial environment. Salto; Landforms: (**e**) Differential erosion caused by action of water in conglomerates of the Ribeirão São Miguel Formation. Vale da Lua, Chapada dos Veadeiros National Park, Goiás; (**f**) Crater edges formed by meteorite impact (elevated hills at the background). Notice the heavy urbanization in the central part of the structure. Colônia Crater, São Paulo. Photos **b-f** by: Maria da Glória M. Garcia

The lithological and structural mosaic that constituted Gondwana was formed by stable, more rigid blocks (cratons), surrounded by highly deformed metamorphic and igneous rocks (orogenic belts) and superposed by the sequences of sediments and sedimentary rocks (cover) (Brito Neves and Fuck 2013). Along the Brazilian margin these terrains are represented by Precambrian igneous and metamorphic

Fig. 3.2 Lithotypes and structures of south-eastern Brazilian rocky shores. Deformed gneisses. (**a**) Paraderivate gneisses showing boudinated mafic intrusions. Santiago Beach, São Sebastião; (**b**) Paleoproterozoic tonalitic gneisses. Jureia Massif, Iguape; (**c**) Leucocratic fine-grained granite dyke in charnockite. Cedro Beach, Ubatuba; (**d**) Locally deformed porphyritic granite showing typical relief. Alcatrazes Island, São Sebastião; (**e**) Mesozoic dyke in granite and gneiss. Ponta do Araçá, São Sebastião; (**f**) Blocks of syenite showing typical weathering patterns. Fome Beach, Ilhabela. Photos by: Maria da Glória M. Garcia

rocks, evidenced in rocky shores, that show registers of the orogenic events that culminated in the amalgamation of the Gondwana Supercontinent (Fig. 3.2a–d). Later, extensive crustal fracturing created fractures through which melted material from the mantle intruded, generating dykes that can now be found in several parts of the coast (Mohriak and Leroy 2012) (Fig. 3.2e). Another important magmatic event is registered by alkaline bodies that occur in the Ilhabela and Cananéia regions (Fig. 3.2f).

3.4 Brazilian Rocky Shores: Bedrock Characteristics, Weathering and Erosional Processes, and Interaction with Biota

Weathering, erosion, transport, and sedimentation processes are part of the rock cycle. All these rock cycle steps are related to the internal and external dynamics of the Earth. Although rocky shores or sand beaches formation is also related to the rock cycle, people generally misunderstand them and their importance to provide biota refuge and life support.

Since the shoreline is an interface or a transition area between continent and ocean, it is affected by both continental and oceanic dynamic processes. Brazilian rocky shores occur because of the outcrop of crystalline rocks such as Precambrian to Mesozoic magmatic and metamorphic rocks. According to Villwock et al. (2005), rocky shores predominate on the south and south-eastern part of the Brazilian coast, between Cabo de Santa Marta (SC) and Cabo Frio (RJ), due to the presence of the Serra do Mar that forms an irregular rugged shore, interrupted by small bays and geologically younger coastal plains (Cenozoic). Thus, the main characteristics of the rocky shore depend on the geological history, lithology, and structure of the rocks that constitute the coast but also on weathering and erosion processes and organisms living there.

In the landscape, rocky shores are fragmented, in the form of boulders, or continuous, such as large bedrocks, as cliffs or exposed vertical rocks. Where rock outcrops are exposed, they are subject to mechanical (or physical), biological, and chemical processes (such as hydrolysis, dissolution, and oxidation chemical reactions) that break down rocks in situ and alter rock properties. Weathering (Fig. 3.3a, b) produces

a b

Fig. 3.3 (**a**) weathering saliencies in mafic dyke due to different lithology and texture: fine-grained basaltic dyke (dark) intruding light colored coarse-grained granite (Saco Grande, Ilha Anchieta, Ubatuba—SP); (**b**) linear grooves produced by weathering processes along the orthogonal set of fractures in granite (Sununga Beach, Ubatuba—SP). Photos by: Christine L.M. Bourotte

sediment and prepares material for erosional and transportation processes. The climatic system (precipitation, temperature, sun radiation, wind, storms) and oceanic factors (tides, waves, sea level rise) also affect these processes along the coast.

It is essential to comprehend the natural processes that control the evolution of these areas since they support these unique ecosystems of the Brazilian coast but also represent spawning areas, birthplaces, refuges, and feeding areas for many species on which local coastal communities depend.

In sedimentary rocks, some layers may present different characteristics and may be more resistant than others, which promotes a differential weathering. Rocks densely jointed or fractured are more susceptible to weathering and erosion since water penetrates more easily and deeply along these discontinuities, altering cohesive rocks (Fig. 3.3a). Rock texture related to size and organization of constituting minerals is also a control factor since the larger ones have a minor contact surface for chemical attack than the smaller ones (Fig. 3.3b).

Thus, weathering processes change the texture and composition of rocks exposed. It breaks down hard and cohesive rocks in situ into weaker and friable rocks, due to the break loose and decomposition of minerals. Depending on the mineral composition of rocks, the weathering agents may act differentially since some minerals are more resistant than others. With time, minerals lose their luster, become more salient, and break loose (Fig. 3.4). As a response to weathering the rocks acquire a rougher texture that provides favorable sites for living organism attachment and refuge.

Chemical and physical weathering occur together and sometimes involve a biological component. The activity of living organisms also contributes to weathering processes. Physical weathering promotes the fragmentation of rocks while chemical weathering alters their chemical composition. Resistant rocks are disintegrated, dissolved, colored, shaped, and broken apart slowly while the process occurs more swiftly for the less resistant rocks. Fragmented rocks become rounded since edges and corners are weathered more quickly than the sides (Fig. 3.5).

Vegetation can break apart rocks when roots of trees or plants grow gradually and continuously into cracks and fractures (Fig. 3.6a). Lichen, fungus, and algae

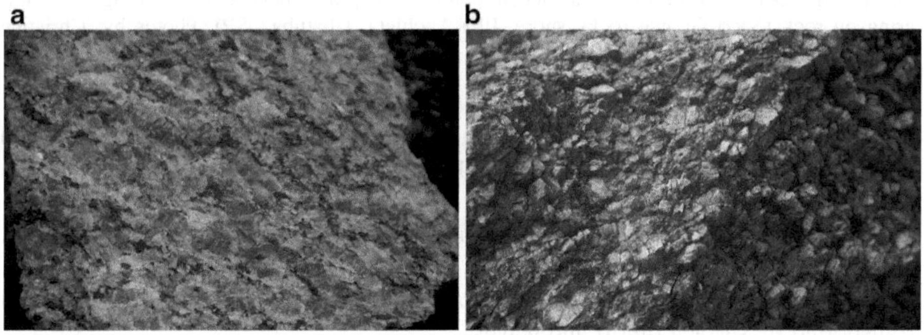

Fig. 3.4 (**a**) Fresh monzogranite (pink feldspar, black mica, gray quartz) and (**b**) weathered monzogranite: minerals lose luster and rock surface is rougher. Photos by: Christine L.M. Bourotte

Fig. 3.5 (**a**) Spheroidal weathering in mafic rocks in which concentric layers of rock break loose and fall off (São Vicente beach, SP). Photo by: Christine L.M. Bourotte; (**b**) spheroidal weathering in granitic rocks in which water moved through the joint promoting chemical weathering that enlarged them gradually rounding the corners and edges of angular blocks (Caraguatatuba—SP). Photo by: Maria da Gloria M. Garcia

Fig. 3.6 (**a**) physical–biological weathering due to root growth in rock cracks and fissures breaking apart the rock; (**b**) yellow and gray lichens colonization promoting chemical–biological weathering on rock (Saco Grande rocky shore, Ilha Anchieta, Ubatuba—SP). Photos by: Christine L.M. Bourotte

also secreting acids that can break down or dissolve minerals, develop cracks and gaps and break the rocks on which they live and get the nutrients they need (Chen et al. 2000) (Fig. 3.6b).

The coastal dynamics promotes erosion and transport: winds generate waves, storms, and surface currents. The gravity of Moon and Sun is responsible for tides. Rocks suffering erosional processes, fragments, and particles are removed and transported by surface runoff and waves.

Erosion differs depending on the structure and lithology of the rocky shore. The rocky shores that are more protected from wave action show more biodiversity than exposed rocky shores because living organisms can more easily fix the former

(Almeida 2008). Waves carrying other rock fragments moving in a turbulent flow can also promote the formation of potholes with different sizes and depths. These cavities are small pools of seawater where living organisms are protected from wave energy and predators (Fig. 3.7a). On the other hand, organisms such as molluscs, sponges, and sea urchins also promote erosion when they dig, drill, scrape, or dislodge particles for fixation, locomotion, or feeding (Naylor et al. 2012). Limpets may be found in minor depressions, crabs in humid crevices, and barnacles attached to the rock (Fig. 3.7b, c). On top of rocks, lichens and terrestrial vegetation such as bromelias, for example, can grow in dryer areas (Fig. 3.7d).

Rocky shores denote an exceptional environment offering sites where organisms live successfully. But at the same time, they are delicate and fragile being susceptible to climatic change, sea level rise, and human activities (as constructions). How can we promote conservation of bio- and geodiversity considering the fragility of natural systems and the advances of urbanization on shorelines? This question will be treated in the next section, oriented by the social learning approach.

Fig. 3.7 (**a**) Potholes are also a refuge for animals such as fish as can be seen at the zoom in the adjacent picture; (**b**) fractures in rocks are occupied by barnacles (*Chthamalus*) and molluscs (*Litorina* and *Brachidontes*); (**c**) tube worm sand construction (*Phragmatopoma*); (**d**) bromelia and lichen covering granite rock where there is no exposure to seawater. Photos by: Christine L.M. Bourotte

3.5 Current Dilemmas of Society and the Role of Environmental Education

Earth system science (ESS) is based on the idea that the Earth can be understood as a set of interacting natural and social systems (Finley et al. 2011). It is emerging rapidly as a discipline that deals with interacting systems. ESS takes the main components of planet Earth—the atmosphere, oceans, freshwater, rocks, soils, and biosphere—and seeks to understand major patterns and processes in their dynamics. The essential ideas describe the general nature of Earth systems, and the analytic concepts describe types of discipline-specific concepts such as the materials, processes, and variables of Earth systems. ESS incorporates a complex view, finding support in the studies that value the environmental issues and the problems of natural risks in terrestrial studies (Carneiro et al. 2005).

Crutzen and Stoermer (2000) emphasize the central role of mankind in geology and ecology by proposing the Anthropocene, the current epoch in which humans and our societies have become a global geophysical force. During the Holocene, accelerating in the industrial period, humankind's activities became a growing geological and morphological force, as recognized by a great number of scientists. The impact of current human activities is projected to last over very long periods. Humans are recognized as geological and geomorphological agents, playing an important role in landscape transformation (Price et al. 2011—see more in Chap. 1).

Environmental education based on Earth system science has the potential to clarify and discuss complex environmental issues, and to contribute to citizens' education towards a sustainable planet. Socioenvironmental complexity, characteristic of densely occupied regions such as the Brazilian coast, has generated "wicked" problems and social conflicts related to the use and occupation of space, causing new dilemmas, inequities, and social inequalities. As cities grow on the coast, and as more and more people choose or require them as places to live, our decisions about urban design and city building will determine the outcomes of long-term challenges related to resilience, sustainability, livability, and justice. Rather than being the essential cause of the global environmental threats we face, cities will be central to success in overcoming these dangers (Maddox et al. 2017). It results in unequal and disproportionate relationships between individuals and nature, characteristic of a predatory society model based on individualistic, competitive, and consumerist values. In addition to this scenario, the absence, inefficiency, or even connivance of public policies is disconnected from the real needs of the community (see more in Chap. 2). The result of this whole process is expressed in the quantitative and qualitative growth of environmental degradation. Geodiversity and biodiversity are also threatened as well as natural resources, with worrisome prospects for local and global societies.

The lack of knowledge and general appreciation of biodiversity and geodiversity contributes even more to the degradation of natural resources. Geological heritage and geodiversity are still little known and not considered as relevant by society. In Brazil, complex biomes are not fully appreciated, but generally considered as obstacles to the development of the economy and civilization, regardless of their ecological importance.

The point is how can we advance science and technology, change human behavior, and influence political will to enable societies to reduce environmental impacts. What knowledge, skills, competencies, and values are needed to cope with new natural, social, political, and economic conditions, and to give shape and meaning to human lives? How can environmental education build upon peoples' own knowledge and skills and offer alternatives to preserve coastal areas, particularly rocky coastlines?

Reflections about these questions bring us challenges and possibilities in how to change human thinking and action in relation to the contemporary environmental issues. Sustainable societies depend on the development of participatory processes, based on dialogic and educational practices, to promote a better understanding of environmental complexity. This implies developing and multiplying socially responsible practices sustained by a new paradigm, in which the relationship between science, society, and public policies is coordinated for the protection of natural resources.

3.6 Social Learning: Concepts and Applications

In the context of this chapter, social learning is defined as "*collective action and reflection that occurs among different individuals and groups as they work to improve the management of human and environmental interrelations*" (Keen et al. 2005). Social learning takes place when divergent interests, norms, values, and constructions of reality meet in an environment that is conducive to learning. This learning occurs at multiple levels including individual, group, or organization, or at the level of networks of actors and stakeholders (Wals 2007; Jacobi 2011).

Reed et al. (2010) argue that to be considered social learning, a process must: "*(1) demonstrate that a change in understanding has taken place in the individuals involved; (2) demonstrate that this change goes beyond the individual and becomes situated within wider social units or communities of practice; and (3) occurs through social interactions and processes between actors within a social network.*"

As a part of environmental education, collaborative practices are an important vehicle to construct a new culture based on dialogue and participation. Its development requires the construction of collective pedagogical spaces, where it is possible to learn and exercise citizenship. Research studies carried out by Santos (2015), Soares (2016), and Mazzucato (2017) point out that social learning frameworks can have a significant contribution for conservation of bio- and geodiversity. The first two studies were carried out in an environment of high complexity, with protected areas and urban occupation, in Guarulhos, State of São Paulo. The last one was developed in a Natural Park, with indigenous and native communities living inside the protected area and with less urbanization, in the coast of São Paulo State. However, in the three studies the methods used were the most appropriate for engaging different stakeholders in environmental issues with successful results and can be applied in other areas.

Considering the complex reality and conflicts of interest inherent in the land use and occupation process in a region of intense urbanization and environmental degradation, Santos (2015) points out that the social learning frameworks have made a significant contribution to the process of understanding the Gold Cycle Geopark, in Guarulhos, metropolitan region of São Paulo city. Through a training course with the participation of teachers, environmental technicians, and public managers, she affirms that the strategies of dialogued learning contributed to the understanding of geoconservation and local sustainability. This understanding presupposes that several actors involved in the process share their ideas and perceptions about the issue/challenge, exploring possibilities and perspectives for joint actions. The course promoted the participation of local leaders and facilitated dialogue among social actors, developing connections between different types of understanding of the problem related to the implementation of the Geopark, thus favoring intersectoral debates as a basis for strengthening cooperation logics.

Soares (2016) analyzed teacher training in Guarulhos, regarding teacher role in the educational process and teaching and learning, as well as promoting dialogue with other stakeholders. A key factor in promoting sustainable protection of natural heritage is the level of awareness of stakeholders, especially local communities (Azman et al. 2010). For this reason, the educational plan was based on the participation of the local community in conservation projects, as a means of protecting the heritage of the place. The concept of a multi-actor platform led to the expansion and the involvement of the different social actors. The experience was based on the use of participatory methodologies that demonstrated effective results in social learning. Through socioenvironmental mapping, the participants investigated different geosites of the region, aiming to understand the geology, geomorphology, biodiversity, soil occupation dynamics, history of the gold cycle, local socioenvironmental problems, and geotourism possibilities. Using role-play, the participants had the opportunity to interpret different social actors and their interests in the region (economic, political, cultural, educational, and touristic). This facilitated understanding of the complex network of relationships in which the geosites are located. With the world-cafe methodology, they experimented to identify the possibilities and limits to geoconservation strategies with dialogue and reflection on the data collected from the local community, as well as to develop collaborative educational and socioenvironmental proposals.

Mazzucato (2017) developed geoconservation strategies in the State Park of Serra do Mar, a conservation area on the coast of São Paulo State, where activities that involve the participation of local communities in the dissemination and valorization of geological heritage in interaction with natural and cultural heritage were promoted. The research identified the perceptions about geodiversity of the different social actors—traditional community, public managers, environmental and NGO monitors, and public school teachers. Relationships between these different perspectives were established based on socioenvironmental diagnosis methodologies and environmental perception with an emphasis on geodiversity. The results point out that the different perceptions of the actors were important to understand and define the geoconservation strategies needed for the place, in a participatory way, contemplating historical, social, and cultural values.

3.7 Discussion and Perspectives

Earth system science is important for the current and future practice of community learning in any kind of environment for reasons that apply to all branches of science education and social practices. It is also allied to sustainability education, which is an area of great and growing interest. The understanding of bio- and geodiversity as part of a complex system is important to promote their conservation. The coastal environment in Brazil is an excellent place to develop projects of environmental education and to promote social learning, as it helps communities and their members respond to the challenge of sustainability in ways appropriate to their situations.

Due to their characteristics, EE can play a critical role in public engagement through clarifying and transmitting the challenges, values, actions, and methods for achieving sustainable, resilient, and just places to live. Decisions about the future uses and conservation necessitate consideration of local characteristics, community engagement in participatory process, and continuous education.

We understand that future directions for conservation in fragile areas, with great interaction of natural factors and process and socioenvironmental complexity as coastal environments, pass through engaging more diverse stakeholders, including underrepresented communities, in a learning collaborative process about the most different aspects of the place. Once understood the relationship between physical, biological, cultural, economic, and environmental aspects of the place, more successful proposes and best practices of sustainable could be carry out.

References

Almeida VF (2008) Importância dos costões rochosos nos ecossistemas costeiros. Cad Ecol Aquát 3(2):19–32

Azman N, Abdul Halim S, Puay Liu O et al (2010) Public education in heritage conservation for Geopark community. Procedia Soc Behav Sci 7:504–511. https://doi.org/10.1016/j.sbspro.2010.10.068

Brito Neves BB, Fuck RA (2013) Neoproterozoic evolution of the basement of the South-American platform. J S Am Earth Sci 47:72–89. https://doi.org/10.1016/j.jsames.2013.04.005

Carneiro CDR, Gonçalves PW, Negrão OBM, Cunha CAL (2005) Ciência do Sistema terra e o entendimento da "máquina" planetária em que vivemos. Geonomos 13(1,2):11–18

Chen J, Blum HP, Beyer L (2000) Weathering of rocks induced by lichen colonization - a review. Catena 39:121–146. https://doi.org/10.1016/S0341-8162(99)00085-5

Crutzen PJ, Stoermer EF (2000) The "Anthropocene". Glob Chang Newsl 41:17–18

Finley FN, Nan Y, Oughton J (2011) Earth systems science: an analytic framework. Sci Educ 95(6):1066–1085. https://doi.org/10.1002/sce.20445

Gray M (2013) Geodiversity: valuing and conserving abiotic nature. Wiley Blackwell, London
Jacobi PR (2011) Aprendizagem Social - Diálogos e Ferramentas Participativas: Aprender Juntos Para Cuidar da Água. IEE/PROCAM, São Paulo
Keen M, Brown SVA, Dyball R (eds) (2005) Social learning in environmental management: towards a sustainable future. Earthscan, London
Maddox D, Nagendra H, Elmqvist T, Russ A (2017) Advancing urbanization. In: Russ A, Krasny ME (eds) Urban environmental education review. Comstock Publishing Associates/Cornell Series in Environmental Education, New York, pp 13–20
Mohriak WU, Leroy S (2012) Architecture of rifted continental margins and break-up evolution: insights from the South Atlantic, North Atlantic and Red Sea-Gulf of Aden conjugate margins. Geol Soc Lond Spec Publ 369:1–40. https://doi.org/10.1144/SP369.17
Mazzucato E (2017) Estratégias de geoconservação no Parque Estadual da Serra do Mar - núcleos Picinguaba e Caraguatatuba (SP). Dissertação (Mestrado em Mineralogia Experimental e Aplicada), Instituto de Geociências, Universidade de São Paulo
Naylor NA, Coombes MA, Viles H (2012) Reconceptualising the role of organisms in the erosion of rock coasts: a new model. Geomorphology 157-158:17–30. https://doi.org/10.1016/j.geomorph.2011.07.015
Price SJ, Cooper AH, Ford JR, Neal C (2011) Humans as major geological and geomorphological agents in the Anthropocene: The significance of artificial ground in Great Britain. Philosophical Transactions of The Royal Society A Mathematical Physical and. Engineering Sciences 369(1938):1056–84. https://doi.org/10.1098/rsta.2010.0296
Reed MS, Evely AC, Cundill G et al (2010) What is social learning? Ecol Soc 15(4):r1. [online] URL: http://www.ecologyandsociety.org/vol15/iss4/resp1/
Santos VMN (2015) Metodologias participativas como práticas de aprendizagem social para sustentabilidade e geoconservação. Relatório de Pós-Doutorado. Instituto de Energia e Ambiente, Universidade de São Paulo
Semken S, Ward EG, Moosavi S, Chinn PWU (2017) Place-based education in geoscience: theory, research, practice, and assessment. J Geosci Educ 65:542–562
Serrano Cañadas E, Ruiz Flaño P (2007) Geodiversity: concept, assessment and territorial application. The Case of Tiermes-Caracena (Soria). Bol Asoc Geógr Esp 45:389–393
Soares DB (2016). Formação continuada de professores em geociências com metodologias participativas: contribuições à educação para geoconservação. Dissertação (Mestrado em Ensino e História de Ciências da Terra), Instituto de Geociências, Universidade Estadual de Campinas
Villwock JA, Lessa GC, Suguio K, Angulo RJ, Dillenburg SR (2005) Geologia e Geomorfologia de Regiões Costeiras. In: Souza CRG, Suguio K, Oliveira AM, Oliveira PE (eds). Quaternário do Brasil. Holos Editora, Ribeirão Preto, pp. 94–113
Wals AEJ (2007) Social learning towards a sustainable world. Principles, perspectives, and praxis. Wageningen Academic Publishers, Wageningen

Chapter 4
Integrating Emerging Pedagogies and Technologies in Environmental and Formal Basic Education

Flavio Augusto de Souza Berchez, Natalia Pirani Ghilardi-Lopes, Sabrina Gonçalves Raimundo, and Antonio Mauro Saraiva

Abstract Emerging pedagogies and technologies are instrumental in adapting educational processes and tools to changes and challenges of the contemporary world. Among them, the global climate changes, together with their associated processes and consequences, bring capital questions for education of all levels, from policymakers to elementary students. This chapter presents the conceptual and practical integration between emerging pedagogies, as outdoor, phenomenon-based education and citizen science, within a transdisciplinary framework, to be applied in formal basic education. Two successful case studies at coastal areas are presented. "Phenomenon-based education" is focused on real-world problem solving and interdisciplinary approaches. Natural environments, the most usual in "outdoor education," offer countless complex and systemic phenomena for analysis, which require not only interdisciplinary collaboration, but also ethical and normative coordination, whence a "transdisciplinary" approach. "Citizen science" is a complementary instrument for studying and comprehending phenomena through actual scientific investigation. All of them are in agreement with "environmental education" principles, which, besides providing an integrated vision of the world that includes

F. A. S. Berchez (✉)
Biosciences Institute and CienTec/USP, University of São Paulo (USP),
São Paulo, São Paulo, Brazil
e-mail: fberchez@ib.usp.br

N. P. Ghilardi-Lopes
Center for Natural and Human Sciences, Federal University of ABC (UFABC),
São Bernardo do Campo, São Paulo, Brazil
e-mail: natalia.lopes@ufabc.edu.br

S. G. Raimundo
Biosciences Institute, Botany Department, University of São Paulo (USP),
São Paulo, São Paulo, Brazil
e-mail: sa.goncalves@usp.br

A. M. Saraiva
Escola Politécnica, University of São Paulo (USP), São Paulo, São Paulo, Brazil
e-mail: saraiva@usp.br

N. P. Ghilardi-Lopes, F. A. S. Berchez (eds.), *Coastal and Marine Environmental Education*, Brazilian Marine Biodiversity,
https://doi.org/10.1007/978-3-030-05138-9_4

all living beings and their environments, also empower people to develop proactive and responsible reactions to social problems. Recent and emerging technologies, including digital media, by contributing to emerging pedagogies, have become important allies for education, especially when considering younger generations, for whom smartphones, tablets, and other devices are essential elements in their lifestyle. Several tools, such as "QR codes," and their alternatives, viz., "radio frequency identification tags" (RFID) and "augmented reality," are discussed, according to how they can be particularly applicable to outdoor education. Efficient "social network" and "educommunication" strategies are complementary educational instruments, fostered by new technologies and digital media, essential to sharing and multiplying highly successful results, and to enabling students, educators, and community members to discuss and solve socio-environmental problems. Our understanding is that these emergent pedagogies and technologies can increase the society's access to formal and environmental education, thus corresponding to relevant tools to sustainability.

Keywords Emerging pedagogies · Phenomenon based education · Outdoor education · Transdisciplinarity · Citizen science · Augmented reality · Social networks

4.1 Introduction

The immense changes in human society that occurred throughout the twentieth century have led to the very complex present-day living conditions prevalent worldwide. Recent advances, as in genetics and computing technology, have speeded up technological, social, and political modifications, to a much higher degree than ever before, thus making everyday life, and perspectives for the foreseeable future, much more dynamic. This is especially true in communications and human relationships (Pantzar 1997). The growing connection among human beings (Harari 2014) has become more accentuated over the last two decades, due to technological development and subsequent by-products, especially Internet and the social media. However, this technological development was not followed at the same speed by philosophical, ethical, and political changes.

Consequently, many hazards have become manifest. Among them, global climate change, together with their associated consequences, comprises not only the most crucial environmental problem currently facing the world, but also a fundamental ethical issue (Gardiner 2004). In this context, and maybe for the first time in history, humankind runs a serious global risk, and thus is challenged to find alternatives and solutions.

Educational processes should adapt to this new world, in such a way as to form individuals and communities capable of dealing with emerging, and huge, challenges. The main concern is twofold: adapt the various publics to the exploitation of new technology and facilitate the efficient access to information and its use (Bayley et al. 2011; Hechter et al. 2012). Developing pedagogies based on novel technology is a way of finding answers to these challenges (Bayley et al. 2011).

In Brazil, although outdoor education activities occur, the use of environmental education (EE) approaches for formal basic education is still incipient. On the other hand, there is a gap between the pedagogies applied in formal education and in EE, which could be greatly enriched by the use of developing techniques and concepts.

In this chapter, we discuss the integrative application of certain emerging pedagogies and technologies to coastal and marine environmental education (CMEE) in Brazil as an instrument for basic formal education, from the present scenario up to potential developments, with special attention on the dialog between outdoor and phenomenological education within transdisciplinary contexts. Efficient "social network," "educommunication," and citizen science are examined as complementary educational tools. Recent experiences using climate change as a transdisciplinary phenomenon will be presented and analyzed as a case study. Some initiatives applying advanced technologies using nature as classroom attributes will also be highlighted.

4.1.1 Emerging Pedagogies

Environmental education concepts are based on the Tbilisi declaration of the First Intergovernmental Conference on Environmental Education (UNESCO 1978), afterwards synthesized by the Rio' 92 Treaty on Environmental Education for Sustainable Societies and Global Responsibility (see more in Chaps. 1 and 2). These proposals consider educational transformation as a key role for achieving both equitable and sustainable societies and ecological conservation.

In many aspects, EE should involve a holistic approach and an interdisciplinary focus (Santos et al. 2018), thereby allowing for an integrated vision of the world, including all living beings and their environments, as well as paving the way for human development of proactive and responsible reactions to social problems.

The concept of interdisciplinarity can be expanded to transdisciplinarity. Max-Neef (2005) discusses the organization of disciplines in education from the isolated approach (disciplinarity) to more complex levels, by involving the horizontal cooperation of disciplines (e.g., pluridisciplinarity), and organizing disciplines on pyramidal hierarchical levels, in which those upper integrate and coordinate those lower. Transdisciplinarity represents the result of coordination at all disciplinary levels, as directed by the knowledge of basic disciplines that represent "what exists" and "how things work," as well as of the upper level, the value level, that indicates "what we should do." The latter is fundamental in the educative process, since distinct ethical views could lead to various interpretations and propositions of solutions.

Hence, improvement of the environmental education process requires developing and practicing alternative pedagogical methods, to thus replace or complement those prevailing. According to Riordan and Klein (2010), methods that include hands-on activities, relevant subject-topics related to the real-life context, and topics that engage students and encourage their active participation are fundamental to EE.

In this sense, certain alternative strategies or emerging pedagogies, as outdoor and phenomenon-based education, may contribute to EE overarching goals. Particularly, the integration between outdoor and phenomenon-based education within a transdisciplinary framework is strongly correlated to EE concepts and practice. Besides, they give a powerful perspective when applied to EE in formal basic education.

Citizen Science and Educommunication are complementary strategies, allowing the participative evaluation, advertising, and multiplication of gains related to the studied phenomenon.

Outdoor learning, often considered as "education in nature," "adventure education," or "experiential education" (Brown 2006), is related to EE through mutual objectives. Natural environments, by offering countless complex and systemic phenomena for analysis, stimulate improvements in problem-solving skills (Sibthorp et al. 2008), while simultaneously providing emotional and cognitive gains that can be associated with physical and mental health, such as improved self-esteem, physical fitness, and a decrease in anxiety and depression (Bogner 1998; Field et al. 2015). This holistic transformation is also related to environmental education, by enabling contemplation of novelties in the current ecological crises, e.g., climate change, since they can reflect on youngsters' perceptions of how such an event might affect their futures (Dunkley 2016). Currently, some European countries are applying this type of methodology, using outdoor learning approaches not only as regards specific environmental education, but also throughout scholastic procedures (Beames et al. 2017; Brown 2009). In the Scandinavian countries, for example, there are schools that include forest-environments in the education process. In these forest schools, all learning processes, especially in early education, are related to nature (Brown 2009; Leather 2018; O'Brien 2009). In this sense, and as alternatives to classrooms, trails or definite natural spaces are more interesting environments that could be used for presentation and discussion of scholar themes while applying a ludic approach.

In "phenomenon-based education," the educative process is related to the interpretation of phenomena. Phenomenon-based teaching, while focused on problem solving and interdisciplinary approaches, is also based on real-world situations, hence providing the starting point for learning. This kind of strategy is particularly interesting since the construction of knowledge is related to the prevailing student context. Examples are place-based learning and problem-based learning, the first more concerned with local interests, and the latter based on problematization as a starting point. Both alternatives are already available and can be used in the EE process.

Citizen science is another possibility for approximating students to the complexity of the real world, through participation in authentic scientific projects and learning by action. Active learning strategies in schools (Grabinger and Dunlap 1995) (1) induce study and research in authentic contexts, (2) encourage a sense of responsibility, initiative, decision-making, and intentional learning on the part of students,

(3) incite collaboration among students and teachers, and further (4) use dynamic, interdisciplinary activities that give rise to complex processes and structures in thinking. Thus, citizen science, also an emerging field, can be an essential environmental education strategy, by reaffirming conservation values and public participation in environmental management (Brasil 1999). This has come about through active citizen–scientist collaboration in data collection, and the monitoring of biodiversity, thereby contributing to the conservation of organisms, by compiling additional information on population structure, spatial distribution, and behavior (Darwall and Dulvy 1996; Delaney et al. 2008; Foster-Smith and Evans 2003; Goffredo et al. 2010).

New pedagogies are based on holistic and integrative perspectives, thereby combining several disciplines and perspectives in problem analysis. However, most experiences are carried out in isolation, in a school, or community. Difficulty in communication frequently precludes outside divulgation, and consequently, enrichment through sharing. Moreover, independent scattered research can result in a loss of time, efforts, and money.

Therefore, the development of efficient social networks and communication strategies, besides being essential for sharing and multiplying successful results, is instrumental in facilitating discussion and problem solving among students, educators, and the community.

Related to this intent, educommunication is a developing discipline aimed at creating and potentializing networks in education, by using traditional and novel media as a form of improving educational experience (Soares 2011). Traditional (e.g., radio) or digital media, such as social networks, sites, or distance learning programs, are used to disseminate audios and videos, transmitting projects, ideas, and solutions. Participation in the creative process can change the students' role from viewers to active actors or educators, by supplying instruments to communicate and disseminate their own opinions and experience.

Although such innovative and emerging approaches and methods are manifold, it is fundamental to place emphasis on significant life experiences during the teaching process (Tanner 1980). The transdisciplinary comprehension of natural phenomena is, therefore, an efficient method for achieving EE principles in basic education and the above-mentioned approaches have the potential to improve such learning in a scholar context.

4.1.2 Emerging Technologies in EE

Recent technologies and novel digital media, together with emerging pedagogies, form important allies in both communication and education, especially when considering younger generations, for whom smartphones, tablets, and other devices have become essential elements. Their use could make education more efficient and entertaining, this becoming even more important when considering nature as an educative space. It is possible that this approach may partially or completely replace traditional education over the medium term. A partial replacement has been occurring and is known as blended learning (Bonk et al. 2012), in other words, a combination of face-to-face and online learning methods.

Among the various computer technologies that could be employed for educational purposes, we can cite mobile devices, such as smartphones (Norris et al. 2011) and tablets, which can be employed to access audios, texts, images, and videos in nature. Smartphones are already being used for educational purposes in schools (e.g., Mendez and Slisko 2013). In addition, technologies have been added to several outdoor activities in situ to promote new ways of learning based on the users' context, with the use, for instance, of videos or photos associated with positioning applications (e.g., Kurti et al. 2008), specially in outdoor programs with predefined routes (Schaal and Lude 2015). However, there are plenty of new strategies coming up every day, since its use enhances the participants motivation and engagement in the activities (Norris et al. 2011). In this way, the educators should always be renewing their knowledge on new technological tools.

In the field, for instance, action could start by the reading of QR codes, or quick response codes, with a mobile device. QR codes are two-dimensional matrix labels, which link to specific webpages on the Internet containing the material to be seen or heard through smartphones. Labels could be distributed in the area to be visited, and attached to certain objects, such as trees, about which the webpage offers information in a variety of media. In the same way, nature components could be used as inanimate teachers, not only giving information, but also proposing puzzles or activities. In Brazil, QR codes are already used for educational purposes (e.g., Nichele et al. 2015), but it is still incipient in environmental education activities. At the University of São Paulo (USP) trails have been conducted using several of those technological implements; and there are also projects in other countries, such as AMULETS at the Växjö University in Sweden, that are based in the same methods (Kurti et al. 2008).

On the other hand, radio frequency identification tags (RFID) use electromagnetic waves at radio frequency to transfer data from an RFID tag attached to the object of interest to a reader containing an information processor (Liu 2009; Liu et al. 2009). This technology has been used, for example, in an aboriginal education course in Taiwan (Chen and Huang 2012). In both cases, contents can be downloaded to the device prior to activity, thus eliminating the need for a Wi-Fi or cellular phone data network.

New technologies, such as augmented reality, are particularly appropriate for the purpose, by facilitating integration with natural elements that could be useful as sources of information, both specific and otherwise. This approach, by humanizing elements from nature, can also lead to affective gains. Augmented reality has become popular from the game "Pokémon GO." In this technology, sound contents and static or dynamic images can be superimposed on the screen of the device, on top of the reality perceived by the user through a camera, thus creating a "new augmented reality," which adds more attraction and content to the scene. At USP, an alternative to the game "Pokémon GO," entitled Bioexplorer, was produced, in which the participants must encounter Brazilian rainforest animals instead of monsters (see a detailed description on item 3 ahead). There are several other possibilities for the use of this technology in education, including environmental education and outdoor learning (Chang et al. 2013; Klopfer and Squire 2008; Wu et al. 2013).

4.2 Global Climate Change Trail: A Phenomenon-Based Learning Activity Case Study

The global climate change trail environmental education protocol was developed by the University of São Paulo (CienTec and IB/USP) and integrates outdoor and phenomenon-based education within a transdisciplinary framework, using educommunication to advertise the final perception of students by themselves.

The phenomenon to be studied, global climate change, is reviewed from a fresh perspective, the photosynthetic process, by using nature as a classroom. Challenges are presented along short trails to groups of five students accompanied by a monitor (Fig. 4.1). The proposal is an understanding of the composition of the Earth's atmosphere and its relationship to the photosynthetic process, the carbon and energy fluxes, and the storage and use of fossil fuels.

Comprehension of the phenomenon depends on knowledge at a disciplinary level (Max-Neef 2005). Apart from chemistry, biology, geography, and geology, other disciplines are also important, such as history, for an understanding of oscillation in fossil-fuel consumption, or mathematics, when considering the balance of compounds in photosynthesis. At the pragmatic level, the main discipline is ecology, through integrating the basic disciplines for a better understanding of the carbon cycle, and the numerous consequences of its increase in the atmosphere. Finally,

Fig. 4.1 Students at work, dealing with the energy flux along food chain (Anchieta State Park, São Paulo, Brazil). Photo by: Flávio A.S. Berchez

normative and ethical disciplines, such as environmental legislation or environmental philosophy (Rozzi et al. 2015), are assets for guiding discussion of the individual, social, and political options involved in facing the problem, thus determining a transdisciplinary approach framework.

A network has been developed for applying this protocol, with the aim of sharing and potentializing efforts (Fig. 4.2). The university is the center for the development of the conceptual and methodological approach of the activity and training of the educators, with execution through dispersed poles, each comprising a protected area (PA) and its respective staff. Besides application in the neighboring schools, the aim is the integration of local students, teachers, and PA monitors through continuous education activities, thereby developing ties that would facilitate the acquisition of skills and stimulate pro-conservation attitudes. The project relies on the continuous training of multipliers at each pole for maintaining a sustainability strategy. Personnel training should involve staff from PAs, fellows, and local volunteers, comprising five certification levels, (1) monitor, (2) monitoring coordinator, (3) tutor, (4) instructor, and (5) examiner, each with specific and growing knowledge and skills.

Action starts at school, with adaptation of content to a specific reality in collaboration with all the teachers, independent of the discipline. Sequentially, student

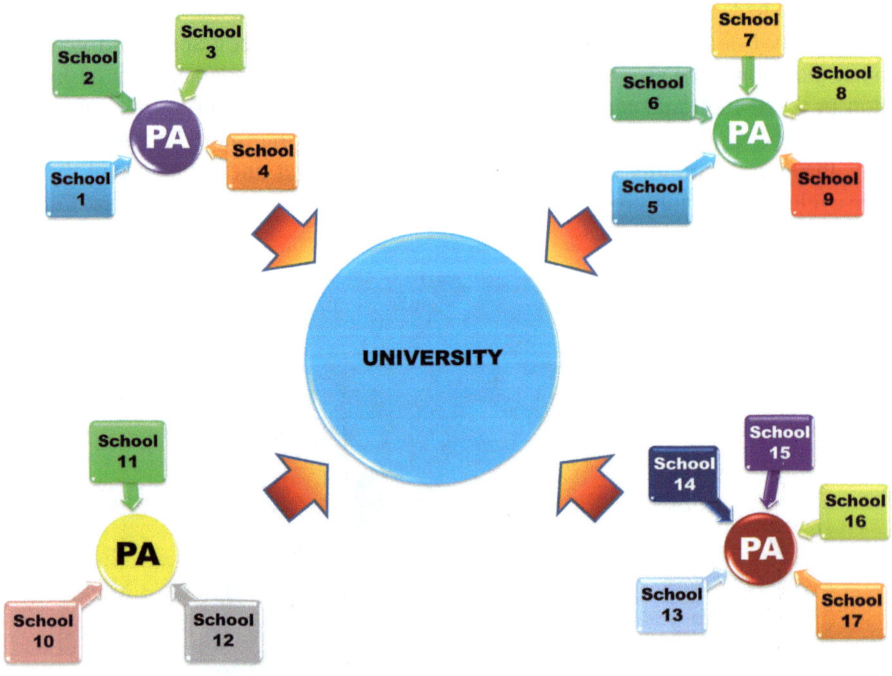

Fig. 4.2 Network coordinating the various actors that collaborate towards applying the "global climate change trail" and other outdoor protocols, structured on a transdisciplinary phenomenon-learning basis. PA = Protected Area

preparation in the classroom is followed by a visit to the PA, with participation in a "global climate change trail" walk. Following post-visit activities that involve group dynamics, exercises, and evaluation testing, students review concepts and knowledge and end up with an evaluation of learning and satisfaction. Each group of five students that took the trail together is stimulated to make a video, so as to compete for the "Annual Global Climate Change" video award. This includes all participant schools. The aim is to encourage students to discuss and communicate their opinion through this kind of media. The results of the first year when this approach was employed, 2016, are summarized in Table 4.1.

Five protected areas were involved, from which two succeeded in carrying out the process until completion. This included staff and teacher training, and student visits, up to development of the final videos. Seven neighboring schools were involved, with 72 teachers having joined the project and received special training. Although 2555 students started, only 1041 continued until completion. From the initial network, four PAs and all the neighboring schools repeated the process during 2017.

Table 4.2 shows the main controversies observed at each level. Lack of interest was a constant reason for failure at all levels, such as the drop-out of 41% of students before video elaboration. On considering the poor conditions and need of incentives for public and environmental education, this could be expected. Nonetheless, the project, although still modest, contributed to promoting and maintaining the interest of part of those involved. Deficiencies in the conceptual basis of students and teachers, although significant, were partially compensated by prior instruction. However, a main problem was related to insufficient student preparation by some of the teachers.

According to school regulations, unforeseen outdoor activities and prior training often required special permission. As the former demands considerable time for organization, especially when without support, working overtime was the solution in most cases. A major problem-limiting participation was the lack of school transportation to the protected areas.

Evasion was a major problem in monitor training. Only 54% of the trained monitors remained active after 2 years of project. Although the proportion was lower among voluntary university students, it was higher among PA members. The main reasons stated were the lack of payment and the need to accumulate 10 h of activity

Table 4.1 Results of the project "global climate change trail" (developed at the University of São Paulo) after 2016, including the protected areas involved, the number of engaged schools by each, the number of trained teachers and of attended students, and whether the project has been continued throughout 2017

Protected area	Schools	Monitors	Teachers	Students	2017
Ilha Anchieta State Park	3	11	19	526	Yes
Juréia-Itatins Ecological Station	4	8	53	515	Yes
Fontes do Ipiranga State Park	0	19	0	1474	Yes
Olympic Streak—USP	0	1	0	0	Yes
Ilhabela State Park	0	5	0	40	No

Table 4.2 Main controversies related to each level of the network of the "global climate change project" (University of São Paulo)

Level	Main controversies
Students	Lack of interest A solid conceptual base for understanding phenomena
Professors	Lack of interest Amount of work involved in outdoor activities Authorization for outdoor activities by school authorities
Schools	Lack of interest Student transportation to protected areas
Monitors	Evasion Lack of financial support
Regional coordinators	Distances Lack of institutional support
Protected areas	Lack of interest Frequent changes in the staff Small number of technicians Inadequate operational and security structures
State authorities	Lack of interest Coordination between units

before the final examination to become a certified monitor. All monitors reaching coordinator status were from remunerated personnel of the PAs: 44% park staff, 28% students with grants, and 28% high school teachers.

The main problems for regional coordinators are the distances between poles, the lack of institutional support for attending training courses, and infrastructure deficiencies, such as transportation and the unavailability of reasonable Internet signal to facilitate meetings by web conference.

There were restrictions both on the number of PA personnel and frequent changes within the staff group. The lack of efficient operational and security structures, as well as adequate equipment, also comprised limiting factors. Even so, only one PA, among all those involved, discontinued the activities.

As reception of the project by state authorities was promising, attempts were made to establish collaboration. At this stage, the main problems were related to formalities, such as delays and an inefficient communication between the upper public administrators and the basic levels managers, such as the school directors.

A preliminary evaluation of results was possible based on content analysis of the video produced by students from the seven schools that concluded the activities. The results indicated the development in the students of a holistic vision of conservation problems, although not always related to global climate changes. Formal scientific research, already under way, is essential for a more profound evaluation of the results (see more on evaluation in Chap. 5).

This type of initiative is not only innovative, but also vital. The transdisciplinary view of processes and questions related to climate change allows a better comprehension and analysis of environmental problems as a whole. The network discussed in this case study is a way to maximize efficiency and lead all actors, from policy-makers to students, to share knowledge, capabilities, and solutions.

Nowadays a good and efficient education on climate change is the greatest environmental challenge (Anderson 2012). However, global climate change education, in and outside schools, is still a topic less valuated both considering mitigation and adaptation goals, even by the IPCC.

4.3 Nature as the Classroom and its Elements as the Teachers: Case Studies Using Emerging Pedagogies and Technologies

In recent initiatives developed by a multidisciplinary group of researchers from the University of São Paulo, natural elements interacted with students through their smartphones. Based on the use of the emerging technologies of QR code, RFID, and augmented reality, as applied through digital devices and technologies, such as smartphones and tablets, digital games, and applications, these strategies are being used to humanize nature and its elements, thereby facilitating audiovisual integration between nature and students or other visitors. Besides affective gains, this approach allows for the use of plants and animals as momentary "teachers," and nature itself as a classroom, thereby improving learning experience.

Fig. 4.3 QR code figure used to access audio and text discussing the relations between urban reserves and cities

This activity is based on a simple use of QR codes, printed on plastic cards and attached to trees, lichens, and the ground (Fig. 4.3). Access to the QR codes through a QR code reader app takes the student to a specific file stored in a website, which includes both image and sound.

The use of RFID tags, encapsulated in the shape of polystyrene nails and inserted into trees, is analogous to that of QR codes. When identified by the mobile phone or tablet via NFC (near field communication), feature that facilitates device data decoding by the chip, the access to the content stored in the RFID tag is retrieved simply by approximation of the apparatus, and without student mediation.

A third technology under development is the use of digital models of virtual objects that appear as augmented reality elements in applications for mobile phones and tablets. In this case, figures of animals, birds, and creatures from Brazilian mythology are scattered throughout a delimited space. When localized by the student, these furnish information, challenges, and exercises through audiovisual information.

The protocols that are being developed using these strategies involve two different phenomena, both linked to conservation. The "plant identity" protocol QR codes placed on trees lead to information on the species and its importance to humans. The game "Bioexplorer" uses figures of animals, birds, and elements from Brazilian mythology to discuss conservation of the Atlantic Rain Forest and its importance (Fig. 4.4).

Specific spaces at protected areas have been delimited to accommodate a specific phenomenon, in an attempt to develop the concept of "classrooms in nature." These spaces can assume various shapes, such as squares or trails where virtual elements, QR codes, or RFID tags can be scattered.

4.4 Perspectives

Improvement of education in Brazil has been slow if compared to developments worldwide, whence the low level in world educational rankings, as PISA (Programme for International Student Assessment - OECD 2016).

For the schools and classrooms enrolled in the global climate change trail project, a preliminary quantitative evaluation of the number of performed outdoor activities, excluded our, was made through a quiz applied to teachers during training activities. The results have shown an average of 0.60 activities.class^{-1}.y^{-1}, with approximately half of the students having had only one extra-class activity throughout the school-year, and the other half, none.

A tendency for a reduction in everyday-life outdoor activities has already been noted among youngsters (Field et al. 2015; McCurdy et al. 2010). Furthermore, school environments are very poor, with most classes still relying on the same instruments used in the beginning of twentieth century, i.e., chalk and chalkboard. This represents a contrast with the outside world, with the diversity and richness of phenomena to be explored, as well as the student's personal technology, namely

Fig. 4.4 A student in a photography with a virtual bird that simulates the Brazilian hawk "Carcará" (*Caracara plancus*). The game "Bioexplorer" also allows the listening of audios on the mobile, with information about the elements, the ecosystems where they would be found, and their vulnerability. Photo by: Flávio A.S. Berchez

smartphones, tablets, and personal computers, which are attractive to them and full of information about everyday life.

Efforts to change this picture are mandatory in promoting, through education, ethical, economic, and technological development. The case studies herein discussed, as well as other experiences beginning in Brazil, are examples of success of integration of new concepts, pedagogies, and technologies for developing novel education strategies. These strategies, together with the link of actors in networks to facilitate application, as well as the preparation of students and professors to deal with them, are ways of fostering formal educational processes, making the adaptation to the new contemporary possibilities and opportunities possible.

References

Anderson A (2012) Climate change education for mitigation and adaptation. J Educ Sustain Dev 6(2):191–206. https://doi.org/10.1177/0973408212475199

Bayley A, Henry T, McBride L, Puckett J (2011) A new era in educational technology: why now? In: Bailey A, Henry T, McBride L, Puckett J (eds) Unleashing the potential of technology in education. The Boston Consulting Group, Boston, pp 6–12

Beames S, Humberstone B, Allin L (2017) Adventure revisited: critically examining the concept of adventure and its relations with contemporary outdoor education and learning. Journal of Adventure Education and Outdoor Learning 17:275–279. https://doi.org/10.1080/14729679.2017.1370278

Bogner FX (1998) The influence of short-term outdoor ecology education on long-term variables of environmental perspective. J Environ Educ 29(4):17–29. https://doi.org/10.1080/00958969809599124

Bonk CJ, Graham CR, Cross J, Moore MG (2012) The handbook of blended learning: global perspectives, local designs. Wiley, Hoboken

Brasil (1999) Lei n° 9.795, de 27 de abril de 1999. Dispõe sobre a educação ambiental, institui a Política Nacional de Educação Ambiental e dá outras providências. Diário Oficial da União, Brasília

Brown M (2006) Adventure education and physical education. In: Kirk D, Macdonald D, O'Sullivan M (eds) The handbook of physical education. Sage, Thousand Oaks, pp 685–702

Brown M (2009) Reconceptualising outdoor adventure education: activity in search of an appropriate theory. J Outdoor Environ Educ 13(2):3–13. https://doi.org/10.1007/BF03400882

Chang HY, Wu HK, Hsu YS (2013) Integrating a mobile augmented reality activity to contextualize student learning of a socioscientific issue. Br J Educ Technol 44:E95–E99. https://doi.org/10.1111/j.1467-8535.2012.01379.x

Chen CC, Huang TC (2012) Learning in a u-museum: developing a context-aware ubiquitous learning environment. Comput Educ 59(3):873–883

Darwall WRT, Dulvy NK (1996) An evaluation of the suitability of non-specialist volunteer researchers for coral reef fish surveys. Mafia Island, Tanzania — A case study. Biol Conserv 78:223–231. https://doi.org/10.1016/0006-3207(95)00147-6

Delaney DG, Sperling CD, Adams CS, Leung B (2008) Marine invasive species: validation of citizen science and implications for national monitoring networks. Biol Invasions 10:117–128. https://doi.org/10.1007/s10530-007-9114-0

Dunkley RA (2016) Learning at eco-attractions: exploring the bifurcation of nature and culture through experiential environmental education. J Environ Educ 47(3):213–221. https://doi.org/10.1080/00958964.2016.1164113

Field SC, Lauzon LL, Meldrum JT (2015) A phenomenology of outdoor education leader experiences. J Exp Educ 39(1):41–44. https://doi.org/10.1177/1053825915609950

Foster-Smith J, Evans SM (2003) The value of marine ecological data collected by volunteers. Biol Conserv 113:199–213. https://doi.org/10.1016/S0006-3207(02)00373-7

Gardiner SM (2004) Ethics and global climate change. Ethics 114(3):555–600. https://doi.org/10.1086/382247

Goffredo S, Pensa F, Neri P et al (2010) Unite research with what citizens do for fun: "recreational monitoring" of marine biodiversity. Ecol Appl 20:2170–2187. https://doi.org/10.1890/09-1546.1

Grabinger RS, Dunlap JC (1995) Rich environments for active learning: a definition. Res Learn Technol 3:5–34. https://doi.org/10.1080/0968776950030202

Hechter RP, Phyne LD, Vermette LA (2012) Integrating technology in education: moving the TPCK framework towards practical applications. Educ Res Perspect 39:136–152

Harari YN (2014) Sapiens: A Brief History of Humankind. HarperCollins, New York

Klopfer E, Squire K (2008) Environmental detectives—the development of an augmented reality platform for environmental simulations. Educ Technol Res Dev 56:203–228. https://doi.org/10.1007/s11423-007-9037-6

Kurti A, Spikol D, Milrad M (2008) Bridging outdoors and indoors educational activities in schools with the support of mobile and positioning technologies. Int J Mob Learn Org 2(2):166–186. https://doi.org/10.1504/IJMLO.2008.019767

Leather M (2018) A critique of forest school: something lost in translation. J Outdoor Environ Educ 21(1):5–18. https://doi.org/10.1007/s42322-017-0006-1

Liu TY (2009) A context-aware ubiquitous learning environment for language listening and speaking. J Comput Assist Learn 25:515–527. https://doi.org/10.1111/j.1365-2729.2009.00329.x

Liu TY, Tan TH, Chu YL (2009) Outdoor natural science learning with an RFID-supported immersive ubiquitous learning environment. Educ Technol Soc 12(4):161–175

Max-Neef MA (2005) Foundations of transdisciplinarity. Ecol Econ 53(1):5–16. https://doi.org/10.1016/j.ecolecon.2005.01.014

McCurdy LE, Winterbottom KE, Mehta SS, Roberts JR (2010) Using nature and outdoor activity to improve children's health. Curr Probl Pediatr Adolesc Health Care 40:102–117. https://doi.org/10.1016/j.cppeds.2010.02.003

Mendez D, Slisko J (2013) Software socrative and smartphones as tools for implementation of basic processes of active physics learning in classroom: an initial feasibility study with prospective teachers. Eur J Phys Educ 4(2):17–24

Nichele AG, Schlemmer E, Ramos AF (2015) QR Codes na Educação em Química. Renote 13(2):1–9

Norris C, Hossain A, Soloway E (2011) Using smartphones as essential tools for learning: a call to place schools on the right side of the 21st century. Educ Technol 51(3):18–25

O'Brien L (2009) Learning outdoors: the forest school approach. Education 3-13 37(1):45–60. https://doi.org/10.1080/03004270802291798

OECD (2016) PISA 2015: Brasil, 2016. Available at: http://www.compareyourcountry.org/pisa/country/bra. Accessed 09 May 2017

Pantzar M (1997) Domestication of everyday life technology: dynamic views on the social histories of artifacts. Des Issues 13(3):52–65. https://doi.org/10.2307/1511941

Riordan M, Klein EJ (2010) Environmental Education in Action: How Expeditionary Learning Schools Support Classroom Teachers in Tackling Issues of Sustainability. Teacher Education Quarterly 37(4): 119–137

Rozzi R, Stuart Chapin III F, Baird Callicott J, Pickett STA, Power ME, Armesto JJ, May Jr. RH (2015) Introduction: Linking Ecology and Ethics for an Interregional and Intercultural Earth Stewardship. In: Rozzi R, Stuart Chapin III F, Baird Callicott J, Pickett STA, Power ME, Armesto JJ, May Jr. RH (eds) Earth Stewardship: Linking Ecology and Ethics in Theory and Practice. Springer International Publishing AG, Cham, pp 1–16

Santos CR, Grilli NM, Ghilardi-Lopes NP, Turra A (2018) A collaborative work process for the development of coastal environmental education activities in a public school in São Sebastião (São Paulo State, Brazil). Ocean Coastal Manag 164:147–155. https://doi.org/10.1016/j.ocecoaman.2017.08.011

Schaal S, Lude A (2015) Using mobile devices in environmental education and education for sustainable development - comparing theory and practice in a nation wide survey. Sustainability 7:10153–10170. https://doi.org/10.3390/su70810153

Sibthorp J, Paisley K, Gookin J, Furman N (2008) The pedagogic value of student autonomy in adventure education. J Exp Educ 31:136–151. https://doi.org/10.1177/105382590803100203

Soares IO (2011) Educomunicação: o Conceito, o Profissional, a Aplicação. Editora Paulinas, São Paulo

Tanner T (1980) Significant life experiences: a new research area in environmental education. J Environ Educ 11(4):20–24. https://doi.org/10.1080/00958964.1980.9941386

UNESCO (1978) Intergovernmental conference on environmental education. Tbilisi (USSR) 74–26 October 1977. UNESCO ED/MD 49, Paris, 101p

Wu H-K, Lee SW-Y, Chang H-Y, Liang J-C (2013) Current status, opportunities and challenges of augmented reality in education. Comput Educ 62:41–49. https://doi.org/10.1016/j.compedu.2012.10.024

Chapter 5
Evaluation of Environmental Education Activities

Rosana Louro Ferreira Silva, Natalia Pirani Ghilardi-Lopes, Sabrina Gonçalves Raimundo, and Suzana Ursi

Abstract In this chapter, we will discuss why and how to evaluate environmental education (EE) activities. Currently, EE activities have been used as a tool for many programs and projects aiming at specifics goals, being fundamental to evaluate them properly and continuously, comparing the results obtained in relation to the proposed aims, verifying the difficulties, redirecting the work, and, finally, measuring the quality of learning. So, evaluation can subsidize both the learner in its development and the educator in the re-dimensioning of its pedagogical practice. A good evaluation begins with the planning phase and continues throughout the application phase. The use of proper methods is essential, either qualitative or quantitative ones, depending on the focus of the questions and evaluation objectives. For that, a evaluation design must consider which method of data collection is going to be used, why it is going to be used, and how to properly analyze the data obtained. In Brazil, there are still a few cases of evaluation of Coastal and Marine EE actions when compared to initiatives in land environments. One of these experiences is described in this chapter: the case study of the Underwater Marine Trail Project (Southeast coast of Brazil).

Keywords Evaluation · Method · Teaching and learning · Pedagogical practice

---

R. L. F. Silva (✉)
Biosciences Institute, Zoology Department, University of São Paulo (USP), São Paulo, São Paulo, Brazil
e-mail: rosanas@usp.br

S. G. Raimundo · S. Ursi
Biosciences Institute, Botany Department, University of São Paulo (USP), São Paulo, São Paulo, Brazil
e-mail: sa.goncalves@usp.br; suzanaursi@usp.br

N. P. Ghilardi-Lopes
Center for Natural and Human Sciences, Federal University of ABC (UFABC), São Bernardo do Campo, São Paulo, Brazil
e-mail: natalia.lopes@ufabc.edu.br

N. P. Ghilardi-Lopes, F. A. S. Berchez (eds.), *Coastal and Marine Environmental Education*, Brazilian Marine Biodiversity,
https://doi.org/10.1007/978-3-030-05138-9_5

5.1 Reflecting About Evaluation Processes

5.1.1 Why Is It Important to Evaluate Environmental Education Activities?

It is plausible to think that every time someone takes time to plan, to develop, and to apply an environmental education (EE) activity, some pedagogical/educational goals might be considered. According to the Belgrade Charter (UNEP, 1975):

> The goal of environmental education is to develop a world population that is aware of, and concerned about, the environment and its associated problems, and which has the knowledge, skills, attitudes, motivations, and commitment to work individually and collectively toward solutions of current problems and the prevention of new ones.

These aims can be detailed in numerous specific objectives. In this context, it is fundamental that the evaluation process of environmental education activities occurs continuously, initiating since the planning phase (Bennet 1989; Catani and Gallego 2009), aiming to boost the process of teaching and learning, in which educator and learner verify and analyze, in a constant process, the meaning of the methods, explanations, and actions in search of knowledge (Justina and Ferraz 2009). Through the evaluation process, the results that are obtained during the joint work of the educator and learners are compared with the objectives proposed to verify progress, difficulties, and redirect the work to the necessary corrections (Libâneo 2017). Moreover, evaluation is one of the most important tools available to educators to measure the quality of learning and offer alternatives for improvement, being an instrument that will subsidize both the learner in its development and the educator in the re-dimensioning of its pedagogical practice (Bennet 1989; Aquino 1997; Stokking et al. 1999; Régnier 2002). Once the educator feels more confident about the difficulties which will be encountered in the teaching and learning process in the educational space, it becomes possible to think more objectively about re-planning environmental actions by taking these difficulties into account.

However, in Brazil and worldwide, the initiatives in environmental education which are accompanied by continuous evaluation are still rare (Guanabara et al. 2009). This fact is worrying, since lack of planning, continuous assessment, and adaptive measures can lead to undesirable results, such as environmental degradation, for instance, as well as misuse of public resources which are destined to carry out these actions.

5.1.2 What Is the Significance of Evaluation in Environmental Education Research?

As well placed in the title of the work of Depresbiteris (2001) called "evaluation in EE: a very delicate relationship," the evaluation process in the field of EE is not simple. Given the complexity of EE and its highlighted objectives, this process needs to incorporate the identification of Awareness, Attitudes, Skills, and Participation, as well as Knowledge, i.e., not only the final results but also the quality of the learning process. The process also involves social, historical, political, and economic aspects, based on human relationships and the grounds of the environmental problems that these aspects reveal, which means that we have multiple perspectives for analysis.

Although the major concern is generally related to evaluation instruments, it is also necessary for the project team to address the following issues: Why evaluate? What to assess? How to evaluate? After the agents of the educational process answer these questions, they will need to think about the appropriate tools to carry out the evaluation.

According to Alba and Gaudiano (1997), the evaluation process is contextualized by three basic characteristics, which break with the traditional notion of the control measures: (1) evaluation is a source of information that allows understanding the pedagogical practice, (2) evaluation is a moment of learning, and (3) evaluation makes it possible to investigate the scope of the process. Although the evaluation also has control objectives (not only of the subjects of the educational action but also of the proponents, materials, and others), it is important that this control be carried out collectively and through democratic structures that allow the participation.

Zint (2012) stresses that "evaluative thinking is essential for supporting the continuous improvement of the environmental education programs". This author did a systematic review of evaluations of EE programs, published in peer-reviewed journals. First, it was noted that the articles were not explicit about their research methodologies, ideologies, or the evaluation approaches in which they were based on. It is important to remember that there are, at least, 15 currents of EE (Sauvé 2010), which have approaches that are very different in aims and values. Another analyzed aspect was that most of the identified articles describe results from evaluations of knowledge (specially concepts) of the participants, although some assessments of attitudes, skills, and behaviors have also been found.

Another important aspect for reflection is the moment of evaluation. There is a consensus that evaluation should permeate all stages of the educational process in environmental education. However, considering that environmental education presents changing as a goal not only at the end of the process but throughout life, authors have mapped research that evaluated the long-term impacts of the projects (Liddicoat and Krasny 2013). In these researches, the evaluation was conducted after months or, in some cases, years of an influential experience. According to Liddicoat and Krasny (2013), there are two possible approaches in this sense: "significant life experiences" and "long-term memory theory." Although the present chapter will not show Brazilian experiences with these characteristics, we emphasize that some behaviors are only expressed years later, "*as a result of supportive and political structures, social norms that differ with age cohort, cumulative experiences and other factors*" (Liddicoat and Krasny 2013 p. 295).

5.1.3 What Can Be Evaluated in Environmental Education Activities?

Since environmental education actions presuppose the interaction between human beings and the environment, we can think that their evaluation can focus on: (1) the subjects or the interaction among the subjects before/during/after the action, (2) the action itself, or (3) the interaction between the subjects and the environment before/during/after the action. We detail each of these focuses ahead.

5.1.3.1 The Subjects or the Interaction Among the Subjects Before/During/After the Action

In this case, the aim is to evaluate the outcomes of the action, or the benefits of the action for the participant's cognitive learning, perception, sensitivity, thinking skills, action skills, attitudes, behaviors, or values, for example. In this matter, it is important that evaluation allows for the identification of unanticipated outcomes (Bennet 1989).

5.1.3.2 The Action Itself

Here, the goal is to verify: (1) the planning—e.g., the adequacy of the objectives regarding the target public, the chosen methodology, the staff involved, and the time available for the implementation; (2) the execution—e.g., if the staff followed the initial planning, if the participants understood the functioning of the action, and if the action was performed within the expected period of time; (3) the outputs—for instance, how many people participated in the action; and (4) the impacts of the action—which changes promoted by the action will be in the long run.

5.1.3.3 The Interaction Between the Subjects and the Environment Before/During/After the Action

The objective here is to evaluate the possible environmental impacts caused by the interaction of the participants with the environment where the action took place.

5.2 Approaches and Methodologies

To understand how to evaluate EE programs or projects and then choose appropriate approaches and methodologies, it is first necessary to know that evaluation process demands the gathering of data to describe what is happening in the program, how it is working, and how well it is going on. With this data available, it is important to learn from them and think how to address the possible problems highlighted by the evaluation.

5.2.1 *Is There a Better Approach to Evaluate Environmental Education Experiences?*

As in other educational processes, qualitative, quantitative, and mixed approaches can be utilized on EE assessments. We can reflect about the discussions that have been carried out about research approaches and apply them to EE evaluation situations. In this perspective, we can consider that, historically, qualitative and quantitative evaluations have been considered opposite approaches. In addition, qualitative methods were developed, many times, in the context of a critique of quantitative strategies. It is true that these approaches possess different epistemology, and, by consequence, the instruments and findings have different origins and natures. However, as postulated by Flick (2014), we can see that both approaches are separate but parallel, with their applicability dependent on the issue and the question to be investigated.

If the evaluators are more concerned with understanding individuals' perception of the word, they probably will choose qualitative approach. In this context, evaluators doubt whether social "facts" exist and question whether a "scientific" approach can be used when dealing with human beings. Qualitative approach not only uses nonnumerical and unstructured data but also, typically, deals with questions and methods which are more general at the start, and become more focused as the evaluation progresses (Bell and Waters 2014). It is important to highlight that qualitative evaluations are not homogeneous.

On the other hand, evaluators are going to use quantitative approaches if their focus is on collecting facts and studying the relationship of one set of facts to another. They use numerical data and, typically, structured and predetermined questions of evaluation, conceptual frameworks, and designs (Bell and Waters 2014). Nowadays, statistical analysis is widely used as an important element of quantitative evaluations.

Mixed approaches can be an appropriate choice in many cases of EE evaluation. For example, evaluators can first conduct a big survey using a questionnaire with multiple choice questions to check short time changes in marine environmental perception after a scuba activity. The answers need to be analyzed by statistical tests. However, the evaluators may would like to carry on a second step of investigation, analyzing long time changes of only a few number of participants. This new goal can be conducted by a case study with qualitative approach, using a variety of methods, like periodic interviews and observations (Flick 2014).

Zint (2012), in her systematic review, stressed that seven studies (the majority in her study) used quantitative evaluative strategies, two used qualitative, and only one used the mixed method. She and other authors emphasize the need to use mixed methods to evaluate the results of environmental education programs, since it better represents the process and results obtained from the actions.

5.2.2 How Can We Measure Outcomes?

It is often difficult to measure the outcomes of an educational activity. For this, projects need be clear about indicators, tools, and means of analysis.

5.2.2.1 Indicators

Indicators can be envisioned as the "flags" that let us know whether we are on the correct path, being the measurable, observable ways of defining our outcomes (Thomson and Hoffman 2017). According to Loureiro (2013), the indicators refer to "a measure, qualitative or quantitative, which informs something specific about a certain aspect of reality." Indicators contribute to the process of evaluation of socio-environmental actions and may vary in accordance with the nature and objectives of the proposed action. Therefore, they can be quantitative (when we want to measure the results objectively and pragmatically, linking goals straightly to the results observed) or qualitative (reflecting those results which are difficult to measure numerically, to better understand process, values, or behaviors, for example). The quantitative representations result in comparable magnitudes and the data numerically provide information about different stages of the teaching–learning process. The qualitative attributes refer to perceptions, representations, values, and behaviors. Indicators are a way of knowing the events that express the parameters adopted in projects, programs, and public policies, allowing an evaluation of the process (Loureiro 2013, p 235). For each outcome, there can be more than one indicator and a different way to design the evaluation (São Paulo 2005).

Indicators should always be articulated to the goals of a project and can be of different types, objective or subjective, and simple (analytical) or complex (synthetic). The objective indicators refer to the quantification of frequency of facts and activities observable in the empirical social reality. On the other hand, the subjective indicators are the descriptions of individuals about their personal and collective perceptions, feelings, wishes, and expectations in relation to certain aspects of reality. Besides, simple (analytical) indicators describe a specific aspect that we wish to measure, without aggregating different sources of information, whereas complex (synthetic) indicators are the synthesis of several simple indicators that express different dimensions of reality (Loureiro 2013). When we think about EE assessment, these indicators can be obtained using evaluation tools, which allow us to identify learning and process elements.

5.2.2.2 Evaluation Tools

There are a variety of instruments that can be used on evaluation processes. We describe some of them below:

a. Observation—an important element of EE assessment. According to Depresbiteris (2001), it can be classified as systematic, in which the observer has predefined objectives and knows which are the aspects that will be evaluated; or unsystematic, identifying casual experiences, registering the largest possible number of information and relating them later, with clear and defined objectives. Some important tools for organizing observation are the field diary, videos, and photographs.
b. Life history—a strategy that invites the different social actors of the process to narrate their experiences throughout life and the process of intervention experienced, indicating not only acquired knowledge but also feelings and engagements regarding participation in the project.
c. Portfolio—in a long process, it is an interesting instrument because it will comprise, for each social actor involved, the set of activities carried out, including annotations, summaries, photos, reports, and reflective essays, among others, allowing to identify reference interpretations.
d. Virtual mural—it has recently been possible to use the support of new technologies for evaluation processes in EE, such as described in Faustino et al. (2017) using the padlet tool. Padlet is a free online platform for digital mural construction available, which can be used free of charge (https://pt.br.padlet.com). In this study, Faustino et al. (2017) used this tool to explain, analyze, and relate the meanings on biodiversity constructed in an individual and collective way in a proposal of continuous education of environmental education.
e. Conceptual map—diagram that represents relationships that a person establishes between concepts, making it possible to identify in what way they structure their knowledge.
f. Questionnaire—a very common instrument in educational research, they may consist of multiple choice, Likert scale (Albaum 1997), or open questions. An instrument that uses more than one type of question is also possible. The instruments need to be validated before their use. Additionally, for a greater dialogue with the international literature, it is desirable to base the contents on relevant theories or approaches, e.g., the Coastal Environment Questionnaire (CEQ—Ursi and Towata 2018), which was specially developed to marine and coastal ecosystems based on Wiseman and Bogner's Model of Ecological Values (Wiseman and Bogner 2003).
g. Interview—an important instrument to educational research and an evaluation tool, which can be structured, semi-structured, or open.
h. Drawing or other artistic production—especially when working with children, drawing is a powerful tool and has been used to evaluate representations of different environmental elements (Schwarz et al. 2007; Pedrini et al. 2010a).

i. Self-assessment—in this process, participation has a central role. The reflective process of the learner immersed in the situation of learning that is happening brings a formative perspective by itself, besides indicating to the educators the relationships that each subject established with the process and how they used or intend to use the elements learned for the daily performance.

Regardless of the instrument or, more appropriately, the set of instruments to be used, the evaluation should be an integral part of the educational process, being formative and not only informative, and used during the whole process, and not just at the end of it. That is why, the directions to develop the instruments to be used might be clear, being part of the design, turning the instruments valid and reliable (Bennet 1989). After knowing the goals, defining indicators, designing how to measure, and developing an appropriate tool to collect data, it is still necessary to analyze the results. Depending on each kind of tool and data, a proper method is indicated.

5.2.2.3 Analyses

Quantitative indicators are generally chosen when the investigator wants to analyze variables (e.g., the comparison of groups considering one or more factors, such as age, schooling, geographical origin, etc.) or indicate the magnitude of outcomes. The data, in this case, can be synthesized through graphics and tables showing the counts or the absolute/relative frequencies and/or abundances of the variable under study. There are several possibilities, and the choice will depend on the nature of the data being collected (Zar 1999). To better understand the data, the calculation of sampling size and normality tests are essential. Since this kind of data allows for the calculation of averages and dispersion, it is possible to perform statistical analysis to test hypotheses of differences among groups and the probability that these differences did not occur by chance. Parametric tests, as analysis of variance (ANOVA), or nonparametric ones, such as Kruskal–Wallis test or permutational analysis of variance (PerMANOVA) (Anderson 2001), are examples of tests that can be performed in this case. Post hoc tests are also essential to properly achieve which groups are different from each other (Fig. 5.1).

On the other hand, the analysis of qualitative data aims at understanding the *big picture* and the main focus is on the process and on the deep description of the phenomenon under analysis. In general, after collecting qualitative data, qualitative evaluation will flow through summarizing and then explaining and structuring the results to get a whole interpretation of them (see review on Flick 2014). The recognition of similarities and differences between groups of data collection can also be performed, and the coding of data is important in this case for the determination of patterns or categories of data. For such, the content analysis (Bardin 2009) is widely used by researchers, in which the information is organized, through a lengthy and detailed evaluation by the researcher, into major and minor categories that represent the cores of the variable under study. Other possibilities include

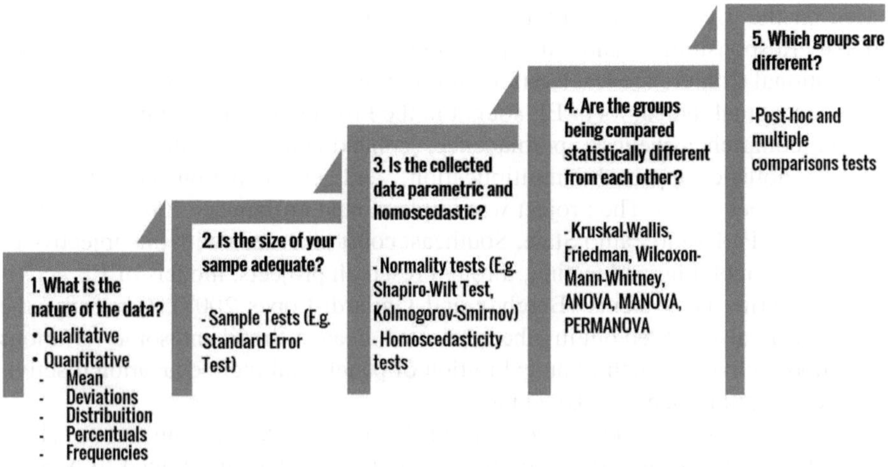

Fig. 5.1 Analysis process of quantitative data. The steps include: (1) definition of the nature of the data (qualitative or quantitative). The description of quantitative data can be done in several different ways; (2) evaluation of sample size; (3) verification of normality and homoscedasticity of data; (4) statistical comparison of the totality of groups; and (5) statistical comparisons between groups

the textual discourse analysis (Moraes 2003), discourse analysis (Gee 2014), and narrative research analysis (Lieblich et al. 1998). Additional approaches such as grounded theory, ethnomethodology, conversation analysis, ethnography, hermeneutic approaches, and many others are reviewed in Flick (2014).

It is important to consider that to obtain good results, well-planned experimental design and adequate tools to collect the data are necessary. For example, in the case of closed questionnaires with affirmatives related to scale options, Alpha Cronbach Test can be used to analyze how consistent the affirmatives are among them (Cronbach 1951). It is also necessary to validate the tools by peer reviews and previous tests. Although it is common to use the qualitative and quantitative approaches separately, both techniques can be complementary to each other. In many cases, mixed methodologies are more adequate to evaluate and achieve a holistic comprehension of the object under study, as exposed above.

5.3 Examples of EE Evaluation in Brazil

We present here examples of EE evaluations carried out within the context of the Underwater Marine Trail Project (see more details on this project in Chaps. 6 and 8). This project offers, essentially, possibilities for activities centered on marine, terrestrial, and virtual tracks, all guided by monitors and with points of interpretation (Berchez and Ghilardi-Lopes 2007). It uses as a theoretical reference the concept of EE from the National Environmental Education Program of Brazil, based on the Treaty on Environmental Education for Sustainable Societies and Global Responsibility and the pedagogical presuppositions of the Tbilisi International Conference Declaration on Environmental Education. Therefore, the main conceptual indicators of EE sought in the Project are transformation, participation, comprehensiveness, permanence, contextualization, ethics, transdisciplinary, holistic approach, multiplication, and emancipation (Berchez and Ghilardi-Lopes 2007). The project was implemented on January 2002 at Anchieta Island State Park (São Paulo State, Southeast coast of Brazil) with the objective of developing, applying, and testing, through research projects, models of EE activities for marine ecosystems (Berchez and Ghilardi-Lopes 2007). Currently, the activities are also carried out in other protected areas in Brazil and some variations of outdoor activities (such as an exhibition of panels and ludic educational activities) held in public schools (Ursi et al. 2010).

The main target audience of the project is basic and undergraduate students, high school teachers, and the protected areas' technicians. They are trained to become agents of multiplication of concepts and knowledge, promoting changing of values and attitudes towards the environment and society. These monitors are carefully trained in short-term theoretical courses, followed by practical internships in Protected Areas (Ursi et al. 2009). Visitors are monitored, consisting mostly of tourists and residents of the northern coast of São Paulo (SP) state, students and teachers of public municipal and state schools of the northern coast of SP, and assistance and recovery entities.

Several scientific works have already been carried out aiming the evaluation and improvement of the Underwater Marine Trail Project (e.g., Berchez et al. 2005, Berchez and Ghilardi-Lopes 2007; Katon et al. 2017; Pedrini and Costa 2007, Pedrini et al. 2008a, b; Savietto et al. 2014; Towata et al. 2013; Ursi et al. 2009, 2010, 2013). In this way, the project is in constant process of transformation, but always remaining faithful to the conceptual indicators of EE on which it was based since its conception. Here, we present two of these study cases.

5.3.1 Study Case on the Environmental Impact of the Tourists During an EE Action: Snorkeling Trail (Berchez et al. 2005)

In 2005, a study was conducted in Anchieta Island Park (southeast coast of Brazil) with the objective of evaluating the possible environmental impact of the participants in one of the activities of the Underwater Marine Trail Project: the snorkeling trail. This trail was installed in a 350-m stretch along of the rocky shore of the Park. It was bounded by a long line of floats running parallel to the coastline at a distance of approximately 5 m far, outlining a safety corridor for the participants of the activity.

Target public consisted of visitors of a range of ages, groups, and educational levels that arrived on the island in tourist-transport schooners. The participants were informed about the activity in a preparatory lecture, in which information was passed on regarding the park and the necessary care to reduce the impact on the biodiversity present on the island, with a view to avoiding touching any species and to come near to the bottom, which could result in touching or in the suspension of sediment. After that, the participants (in groups of five or less) were conducted to the guided interpretative snorkeling trail. All the project activities last approximately 1 h. Two guides accompanied the participants, and one of them was responsible for evaluating the snorkeling of the participants. The observations of this latter guide made it possible to identify infringements of the initial recommendations stressed during the initial lecture, such as the involuntary touching of the substrate (fins), the involuntary touching of the substrate (hands), the descent till the bottom, the wandering from the group, the lack of interest, the voluntary touching of the substrate, and the suspension of bottom sandy–muddy sediments.

The touching of fins on organisms promoted breakings (for example, in the bryozoan *Schizoporella* sp.) or removals from the substrate (for example, of the ascidian *Phallusia nigra*, the crinoid *Tropiometra carinata*, or the seaweeds *Padina gymnospora* and *Dichotomaria marginata*). After this evaluation, some alterations in the trail were implemented. For example, to reduce the wandering of the participants from the group, a rectangular format for the raft support was chosen and it was used to gather the participants during the activity (Fig. 5.2a, b). Also, new models of activities were created, focusing on people with less or no snorkeling abilities, so that no previous experience in diving was needed: (1) a monitored visit to a natural aquarium (tide pool) where organisms are observed through an underwater observation device and the participants simply walk through a trail installed in the tide pool (Fig. 5.2c, d) and (2) a self-guided indoor activity with the use of educational posters (Fig. 5.2e, f).

5.3.2 Study Case on Environmental Perception and Social Representation of Teachers in a Formative Experience During EE Actions (Katon et al. 2017)

One of the difficulties to protect natural environments relies in the differences in perceptions of the values and importance of those between individuals of different cultures or different groups. We consider that understanding Environmental Perception of local population is an important step towards biodiversity conservation. We understand environmental perception as the relationship that people establish with the environment in which they are inserted, which occurs through perceptual and cognitive mechanisms (Whyte 1977). Some authors (e.g., Ghilardi-Lopes et al. 2015) highlighted the strong relationship between Environmental Perception and Social Representations as well.

Fig. 5.2 (**a/b**) Rectangular raft for support of participants during the Underwater Marine Trail (snorkeling model), (**c**) observation device created for those participants of the Underwater Marine Trail with no snorkeling abilities, (**d**) use of observation device in the natural aquarium (participants simply walk into the water and use the device to observe the environment), (**e**) indoor activity simulating the Underwater Marine Trail with the use of educational posters, and (**f**) children performing the indoor activity with educational posters. Photos by: Natalia P. Ghilardi-Lopes

In 2012, a study was conducted aiming to investigate the environmental perception of a group of 11 teachers from a public school (located in Ubatuba city, Southeast coast of Brazil) who participated in a formative experience in environmental education activities. This study presented as specifics goals: (1) to verify whether and how the participation in educational activities influenced the environmental perception of such teachers, and (2) to investigate the conceptions that teachers have about the possibilities and challenges of teaching practice in addressing marine and coastal environment from the subsidies provided by the training experience.

The teachers participated in a variety of Underwater Trail Project activities that were developed along 2 days at Serra do Mar State Park and at Ilha Anchieta State Park, both in Ubatuba city. Different data collection tools were applied at the beginning and at the end of the experience (such as questionnaires and writing of texts), besides the transcripts of the audiovisual material related to collective reflections made at the end of each day. The evaluation of the data collected was done by open categorization and Bardin's Content Analysis technique (Bardin 2009).

We noticed that the participation on environmental education activities influenced, even subtly, teacher's environmental perception. This effect was more evident in specific aspects, such as, for example, in relation to the notion of the biodiversity of these environments. Although not all the teachers expressed the perception in relation to negative impacts on the environment during the activities, when such impacts were perceived, they were closely related to the rocky shore organisms, specifically to their trampling. We also found interesting aspects about the social representations of those teachers which, even after participating in the experience, remained anthropocentric. About the possibilities and challenges of teaching practice in addressing marine and coastal environments, we noticed that for subject teachers of this research, field activities are the most significant to work with such a theme. In addition, we have noticed the great importance that the exchange of experiences among peers, as provided by the experience, present in the composition of teaching practices.

5.4 Final Considerations

According to Zint (2012), "evaluative thinking is essential for supporting the continuous improvement of the environmental education programs that are so desperately needed to help address environmental challenges." In this chapter, we intended to show the complexity of the evaluation in environmental education and to guide educators on its different dimensions, which were synthesized in Fig. 5.3.

We would like to encourage the educators and researchers in environmental education to think about these different dimensions of evaluation as a key part of their programs. We presented two examples to explore some practical possibilities of evaluations in marine environmental education programs. However, we emphasize the need for further studies using long-term assessments.

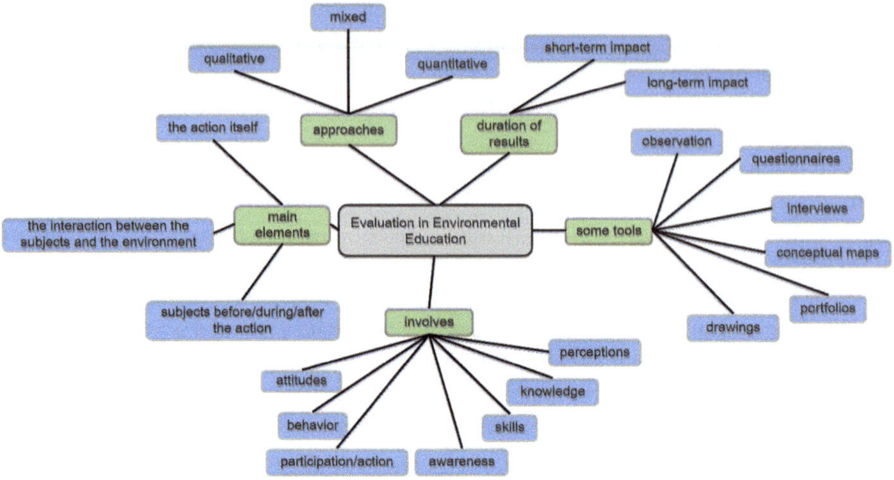

Fig. 5.3 Conceptual map representing the different dimensions of the evaluation process that need to be incorporated in environmental education (using Lucidchart—https://www.lucidchart.com/pages/pt/criador-de-mapas-conceituais)

References

Alba A, González Gaudiano E (1997) Evaluación de Programas de educación ambiental: experiencias en America Latina y el Caribe. Universidad Nacional Autónoma de México, Coyoacán

Albaum G (1997) The likert scale revisited. Mark Res Soc J 39(2):1–21. https://doi.org/10.1177/147078539703900202

Anderson MJ (2001) A new method for non-parametric multivariate analysis of variance. Aust Ecol 26:32–46. https://doi.org/10.1111/j.1442-9993.2001.01070.pp.x

Aquino JG (1997) Erro e fracasso na escola: alternativas teóricas e práticas. Summus, São Paulo

Bardin L (2009) Análise de conteúdo. Edições 70, Lisboa

Bell J, Waters S (eds) (2014) Doing your research project: a guide for first-time researchers. McGraw-Hill Education, New York

Bennet D (1989) Four steps to evaluating environmental education learning experiences. J Environ Educ 20(2):14–21. https://doi.org/10.1080/00958964.1989.9943026

Berchez F, Carvalhal F, Robim MJ (2005) Underwater interpretative trail: guidance to improve education and decrease ecological damage. Int J Environ Sustain Dev 4(2):128–139. https://doi.org/10.1504/IJESD.2005.007235

Berchez F, Ghilardi-Lopes NP (2007) Projeto Trilha Subaquática: sugestão de diretrizes para a criação de modelos de Educação Ambiental em unidades de conservação ligadas a ecossistemas marinhos. OLAM Ciência & Tecnologia 7:181–209

Catani DB, Gallego RC (2009) Avaliação. UNESP, São Paulo

Cronbach LJ (1951) Coefficient alpha and the internal structure of tests. Psychometrika 16:297–334. https://doi.org/10.1007/BF02310555

Depresbiteris L (2001) Avaliação da Aprendizagem na Educação Ambiental: uma relação muito delicada. In: Sato M, Santos JE (eds) A Contribuição da Educação Ambiental à Esperança de Pandora. RIMA, São Carlos, pp 531–557

Faustino MT, Roberto ECO, Silva RLF (2017) Utilizando um mural digital para investigar significados da biodiversidade apresentados por professores(as). Ensenanza de las ciencias (nº Extraordinario):3467–3471

Flick U (ed) (2014) An introduction to qualitative research. Sage, London
Gee JP (ed) (2014) An introduction to discourse analysis: theory and method. Routledge, London
Ghilardi-Lopes NP, Turra A, Buckeridge MS et al (2015) On the perceptions and conceptions of tourists with regard to global environmental changes and their consequences for coastal and marine environments: a case study of the northern São Paulo state coast, Brazil. Mar Policy 57:85–92. https://doi.org/10.1016/j.marpol.2015.03.005
Guanabara R, Gama T, Eigenheer EM (2009) Contribuições para a construção de uma matriz para avaliação de projetos de educação ambiental. Educação e Pesquisa, São Paulo 35(2):399–411. https://doi.org/10.1590/S1517-97022009000200012
Justina LAD, Ferraz DF (2009) A Prática Avaliativa no Contexto do Ensino de Biologia. In: Caldeira AMA, Araujo ESNN (eds) Introdução à didática da Biologia. Editora Escrituras, São Paulo, pp 233–246
Katon GF, Towata N et al (2017) Social representation of marine and coastal environments: a perspective of teachers from southeast coast of Brazil. Conexão Ci 12(2):298–304
Libâneo JC (2017) Didática. Cortez Editora, São Paulo
Liddicoat K, Krasny M (2013) Research on the long-term impacts of environmental education. In: Stevenson RB, Brody M, Dillon J, Wals AEJ (eds) International handbook of research in environmental education. Routledge, New York, pp 289–297
Lieblich A, Tuval-Mashiach R, Zilber T (eds) (1998) Narrative research: reading, analysis, and interpretation. Applied social research methods series. SAGE, Thousand Oaks
Loureiro CFB (2013) Indicadores - Meios para a avaliação de projetos, programas e políticas públicas em educação ambiental. In: Ferraro Junior LA (ed) Encontros e Caminhos: Formação de Educadoras(es) Ambientais e Coletivos Educadores – Volume 3. MMA/DEA, Brasília, pp 235–243
Moraes R (2003) Uma tempestade de luz: a compreensão possibilitada pela análise textual discursiva. Ciência & Educação 9:191–211. https://doi.org/10.1590/S1516-73132003000200004
Pedrini A, Costa EA, Ghilardi N (2010a) Socially vulnerable children and pre-adolescents environmental perceptions in education projects. Ciência & Educação 16(1):163–179. https://doi.org/10.1590/S1516-73132010000100010
Pedrini AG, Costa C (2007) Efeitos ambientais da visitação turística em áreas protegidas marinhas: estudo de caso na Piscina Natural Marinha, Parque Estadual da Ilha Anchieta, Ubatuba, São Paulo, Brasil. OLAM Ciência & Tecnologia 7:678–696
Pedrini AG, Costa C, Silva VG et al (2008b) Gestão de áreas protegidas e efeitos da visitação ecoturística pelo mergulho com snorkel: o caso do Parque Estadual de Ilha Anchieta (PEIA), São Paulo, Brasil. Revista Eletrônica do Mestrado em Educação Ambiental 20:1–20
Pedrini AG, Dutra D, Robim MJ, Martins SL (2008a) Gestão de Áreas Protegidas e Avaliação da Educação Ambiental no Ecoturismo: Estudo de Caso com o Projeto Trilha Subaquática - Educação Ambiental nos Ecossistemas Marinhos no Parque Estadual da Ilha Anchieta, São Paulo. OLAM Ciência & Tecnologia 8:31–55
Pedrini AG, Messas T, Pereira ES et al (2010b) Educação Ambiental pelo Ecoturismo numa trilha marinha no Parque Estadual da Ilha Anchieta, Ubatuba (SP). Revista Brasileira de Ecoturismo 3:428–459
Régnier JC (2002) A Auto-avaliação na Prática Pedagógica. Rev Dia Educacional 3(6):1–16
São Paulo (2005) Manual para Elaboração, Administração e Avaliação de Projetos Socioambientais. Secretaria do Meio Ambiente (SMA/CPLEA), São Paulo
Sauvé L (2010) Educación científica y educación ambiental: un cruce fecundo. Enseñanza de las Ciencias 28(1):5–18
Savietto SF, Katon GF, Towata N et al (2014) Ambientes Marinhos e Costeiros: qual a percepção de estudantes de escolas do litoral norte de São Paulo. Revista da SBEnBio 7:6746–6757
Schwarz ML, Sevegnani L, Andre P (2007) Representações da mata atlântica e de sua biodiversidade por meio dos desenhos infantis. Ciência & Educação 13(3):369–388. https://doi.org/10.1590/S1516-73132007000300007
Stokking K, van Aert L, Meijberg W, Kaskens A (1999) Evaluating environmental education. IUCN, Cambridge

Thomson G, Hoffman J (2017) Measuring the success of environmental education programs. Canadian Parks and Wilderness Society, Ottawa

Towata N, Katon GF, Berchez FAS, Ursi S (2013) Ambiente marinho, sua preservação e relação com o cotidiano: influência de uma exposição interativa sobre as concepções de estudantes do Ensino Fundamental. Enseñanza de las Ciencias (volume extra): 3554–3559

Ursi S, Ghilardi-Lopes NP, Amancio CE et al (2010) Projeto Trilha Subaquática virtual nas escolas: proposta de uma atividade didática sobre o ambiente marinho e sua biodiversidade. Revista da SBEnBIO 3:3821–3829

Ursi S, Towata N (2018) Environmental perception about marine and costal ecosystems: evaluation through a research instrument based on model of ecological values. Problems of education in the 21st century 76(3): 393–405

Ursi S, Towata N, Katon GF, Berchez FS (2013) Influência de exposição interativa sobre ambiente marinho e sua biodiversidade nas concepções de meio ambiente de estudantes do Ensino Fundamental. Enseñanza de las Ciencias (volume extra): 3575–3580

Ursi S, Towata N, et al (2009) Concepções sobre Educação Ambiental em curso de Formação para educadores do projeto Ecossistemas Costeiros (Instituto de Biociências - USP). In: Anais do VII Encontro Nacional de Pesquisa em Educação em Ciências, Universidade Federal de Santa Catarina, Florianópolis, pp 8–13

UNEP (1975) The Belgrade Charter: a framework for environmental education. In: International Workshop on Environmental Education, Belgrade, 1975. 4p.

Whyte AVT (1977) Guidelines for fields studies in environmental perception. UNESCO/MAB, Paris

Wiseman M, Bogner FX (2003) A higher-order model of ecological values and its relationship to personality. Personal Individ Differ 34:783–794. https://doi.org/10.1016/S0191-8869(02)00071-5

Zar JH (ed) (1999) Biostatistical analysis. Prentice-Hall, Upper Saddle River

Zint M (2012) Advancing environmental education programs: insights from a review of behavioral outcome evaluations. In: Brody M, Dillon J et al (eds) International handbook of research in environmental education. Routledge, New York, pp 298–309

Part II
Implementation of Coastal and Marine Environmental Education in Brazil

Chapter 6
Emblematic Coastal and Marine Environmental Education Projects in Brazil

Alexandre de Gusmão Pedrini, Laura Pioli Kremer, Daniel Shimada Brotto, and Valeria Rocha França da Silva

Abstract Aiming to briefly present and discuss examples of Coastal and Marine Environmental Education (CMEE) in Latin American Countries, six emblematic Brazilian projects were selected according to six different categories: *(a) Traditional Populations as social actors, represented by Tamarzinhos Project; (b) Charismatic Species, Golfinho Rotador Project; (c) Rocky shore Ecosystems, Underwater Marine Trail Project; (d) Simulacrum, Paraíba Marine Aquarium; (e) Marine Ecotourism Product, ProMar Project; and (f) Marine Sciences Courses in Public Schools, graduate Program in Marine Sciences applied to teaching.* Although these successful projects target at diversified publics and present different methodologies, their main objective is to promote changes in values, attitudes, behaviors, and postures to transform and emancipate people. The analysis of the projects allowed to conclude that: (a) they cannot count on medium and long-term financial support, but when they have funding it proved to be quite effective to propitiate good results; (b) the teams involved in the projects are not always multidisciplinary in face of their high-cost maintenance; (c) all projects present regular qualitative and/or quantitative evaluations; (d) the main difficulties were always related to financial support, and funding obtained from private sector has been the solution sought to face these difficulties. Despite all the difficulties found, the CMEE projects analyzed have been satisfactorily developed, mainly due to the commitment of their idealizers.

A. G. Pedrini (✉) · D. S. Brotto
Vegetal Biology Department, Rio de Janeiro State University (UERJ), Maracanã, Rio de Janeiro, Brazil
e-mail: pedrini@uerj.br

L. P. Kremer
Campus Itajaí, Instituto Federal de Santa Catarina (IFSC), Itajaí, Santa Catarina, Brazil
e-mail: laura.kremer@ifsc.edu.br

V. R. F. Silva
Projeto Tamar, Fundação Pró-Tamar, Salvador, Bahia, Brazil
e-mail: valeria@tamar.org.br

N. P. Ghilardi-Lopes, F. A. S. Berchez (eds.), *Coastal and Marine Environmental Education*, Brazilian Marine Biodiversity,
https://doi.org/10.1007/978-3-030-05138-9_6

Keywords Coastal and Marine Environmental Education · Coastal and Marine Environmental Education Projects · Charismatic species · Transformative and Emancipatory Environmental Education · Project evaluation

6.1 Introduction

Although Coastal and Marine Environmental Education (CMEE) presents several possibilities in Latin American countries, Research in Environmental Education is a field that has not yet been explored in depth. The few initiatives come from Mexico and Brazil, where there is an attempt to institutionally promote educational research in the area. In other countries, the effort consists in carrying out research projects by a few researchers and some isolated institutions that, under precarious conditions, develop such investigations (González-Gaudiano and Lorenzetti 2009).

In addition, there is an important limitation related to the misconception about the own meaning of EE, which is quite common even among its promoters (Pedrini and De-Paula 2011) (see more on the definition of EE used here in Chap. 2). For example, in scientific popularization (SPop), EE is often treated erroneously (Hadel 2010), sometimes being used as a synonym for Environmental Interpretation (EI) or biology classes (Pedrini et al. 2010) and even, ecology or general science (Pedrini 2010). There are many examples of the confusion generated due to this lack of conceptual delimitation, as marine trails used in SPop projects being included in EE bibliographic databases or, on the other hand, projects aiming at the effective inclusion of EE in scuba diving briefings not always being considered as CMEE. Due to this situation, the retrieval of scientific papers from bibliographic databases, made by any reviewer on this subject, will certainly hamper, underestimating the subject and perpetuating serious mistakes that will be propagated in future inventories of EE projects, weakening initiatives on this subject.

Specifically, in Brazil, even though several efforts were made by governmental entities, there is a lack of a complete and reliable record and analysis of EE projects, consisting solely on inventories comprising specific periods of time or bibliographic units (BU), such as academic theses or dissertations (Reigota 2007). Due to the extremely diversified geographic and social contexts in the country, the "state-of-the-art" researches often resulted in partial and unstandardized analyses. More recently, the EArte project (2018) made available a database of Brazilian academic dissertations and theses on EE produced between 1981 and 2016, including 4205 documents, 11 of which having "marine" or "coastal" as keywords.

If we consider EE projects in Brazilian schools, an inventory was made by the Environmental Education Coordination Committee (EECC) of the Minister of Education and Culture (COEA 2001). Among 30 subject themes found in this survey, just sandbank/dunes, mangroves, and aquatic ecosystems were mentioned as CMEE and it seems that the related EE activities undertaken solely consisted of occasional or periodical short courses, beach cleaning, and planting campaigns. These kinds of EE activities pedagogically contradict those stated by Meadows (1989), and the recommendations stated by the National Environmental Education

Program (Brasil 2005—see more in Chap. 2). Also, there isn't a continuous evaluation of the projects under way (see more on evaluation in Chap. 5).

Although the great majority of EE initiatives in Brazil are short-termed, it is possible to identify some that present a sound theoretical basis and are persisting over time. This chapter aims to present six of these emblematic and successful Brazilian CMEE experiences, characterizing and describing them according to EE theoretical references, aiming to present some cues for the management of EE projects and furnishing some basis for the settlement of patterns to direct future initiatives.

6.2 Selection Criteria and Methodology for Data Gathering

The six emblematic projects were selected to explore a variety of contexts and approaches of CMEE along the Brazilian coast based on the following criteria: (a) Traditional Populations as social actors—here represented by Tamarzinhos Project, (b) Charismatic Species—here represented by Golfinho Rotador Project, (c) Rocky shore Ecosystems—here represented by the Underwater Marine Trail Project, (d) Simulacrum—here represented by the Paraíba Marine Aquarium, (e) Marine Ecotourism Product—here represented by ProMar Project, and (f) Marine Sciences Courses in Public Schools—here represented by the graduate (*lato sensu*) Program in Marine Sciences applied to teaching.

A clear and objective questionnaire was created aiming to gather information from the above specified projects. The questionnaire was sent by email to the coordinators of each project, to be filled by them or a person in charge. It was intended that each questionnaire would attain the required data to draw a clear and as complete as possible profile of each project. The first requested topic of the questionnaire was a brief historical description of the project, followed by the questions aiming to specify the following topics: (a) project maintainer institution and external financing source; (b) conceptual characteristics of the projects, including their target public, team composition, objectives, methods, evaluation strategies, and pedagogical products; (c) encountered difficulties; and (d) adopted solutions.

6.3 The Projects and Their Institutional Financing

The selected projects showed a great diversity of maintainer institutions (Table 6.1). There were federal agencies such as Chico Mendes Biodiversity Institute (ICMBio) and the Federal Institute of Santa Catarina (IFSC) or state institutions as the State University of São Paulo (USP). Paraíba Aquarium was maintained by private companies, while the Tamarzinhos Project relied on the sponsorship and sustainability strategy through the Pro-Tamar Foundation.

Table 6.1 Characteristics of the selected projects according to their typological classification, maintainers, and bibliographical references that detail the project

Typology	Project	Maintainer institution	Reference
Traditional Populations as social actors	Tamarzinhos Project	Pro-Tamar Foundation	Silva et al. (2015)
Charismatic Species	Golfinho Rotador Project	Chico Mendes Biodiversity Institute (ICMBio)	Silva Junior et al. (2010)
Rocky shore Ecosystems	Underwater Marine Trail Project	University of São Paulo Biosciences Institute (IB-USP)	Berchez et al. (2005)
Simulacrum	Paraíba Aquarium	Paraíba Aquarium	Aquário Paraíba (2018)
Marine Ecotourism Products	ProMar project	Nongovernmental Organization ProMar	Rhormens et al. (2017)
Marine Sciences Courses in Public Schools	Graduate Program in Marine Sciences applied to teaching	Federal Institute of Santa Catarina (IFSC)	Berchez et al. (2016)

6.3.1 *Tamarzinhos Project*

Tamarzinhos Project is a Marine Environmental Education course that promotes ocean literacy (see Chap. 1) associated with marine conservation among children, being developed at the Tartarugas Marinhas Project (TAMAR; Marine Turtle Project) based at Praia do Forte (Bahia State) and lasting one year. It seeks to understand the perception that these children have about the environment in which they live, using Marine Turtles (Fig. 6.1) as a guiding element in the activities focused on the affirmative view of local culture, environmentally sustainable practices, and preservation of the various surrounding ecosystems. It was created in 1996 and has been evolving since then. Each edition includes a group of 20 children of the local community, from 10 to 14 years old, selected by the criteria of being enrolled at a school, having territorial ties with the locality and being residents from Praia do Forte surrounding communities (Pegas et al. 2012).

6.3.2 *Golfinho Rotador Project*

The Golfinho Rotador (Spinner Dolphin) Project was created in 1990 aiming at the conservation of marine biodiversity, using the charisma of dolphins and the ease of presenting ocean wildlife at Fernando de Noronha (FN) island, a marine protected area in Northeastern Brazil. It acts by empowering, raising awareness and sensitizing entrepreneurs, service providers, students, public agents, and tourists to adopt

Fig. 6.1 Children attending an activity from Tamarzinhos Project, Bahia State, Brazil. Photo by: Valeria R. F. Silva

sustainable practices in their daily lives. It was sponsored by its own coordinators and by the Brazilian public oil company. The project coordinator participate in the implementation of four national action conservation plans and also in the process of evaluating the conservation state and the network of stranded aquatic mammals in Brazil, in addition to the formulation of local public policies through the FN advisory councils, contributing to develop a broad research program for the conservation of cetaceans.

6.3.3 Underwater Marine Trail Project

The Underwater Marine Trail Project started in 2001 at Ilha Anchieta State Park, a marine protected area in São Paulo State, and since then models of activities with a standardized conceptual basis have been developed, such as a snorkeling trail, a SCUBA diving trail, a tide pool visitation, a kayaking route in the mangroves, a terrestrial trail, and, finally, in 2006, a theoretical trail, constituted of interactive posters (Fig. 6.2). The project was envisioned with the objectives of: (1) developing, applying, and testing, through research projects, environmental education standardized models, in search for standards for CMEE activities, aiming to contribute to structuring similar projects in other locations; and (2) transforming and favoring the participants in knowledge, affectivity, and capacities, promoting a change of expectations and world view, and awakening them to recreational and professional activities related to the marine environment and the importance of their ecosystems.

Fig. 6.2 Snorkeling and SCUBA trails of the Underwater Marine Trail Project, São Paulo State. Photo by: Alexandre de G. Pedrini

Fig. 6.3 Marine interpretive trail of ProMar project. Photo by: Marta S. Rhormens

6.3.4 ProMar Project

The ProMar Project developed a model of marine ecotourism in a reef ecosystem at the Environmental Protection Area of Tinharé and Boipeba, a coastal protected area on the northeast coast of Brazil, which presents excellent conditions for the implementation of ecotourism products such as marine interpretative trails (Fig. 6.3). A 320-m-long trail was installed at Tassimirim beach, which presents diversified marine reefs. Implementation methodology was based on seven steps, starting with community mobilization. By this method, local traditional people were trained as conductors to offer an effective marine ecotourism product to tourists. About 89% of the participants who tested the product evaluated it as "excellent"; 76% of participants would accept to pay \$17.00–\$33.00 for the product; and more than 79% of the local residents and entrepreneurs found the initiative excellent. The product was tested for 10 days and generated a total income of US\$ 433.00. It was concluded that the CMEE product showed efficacy and economic/environmental sustainability. It was well accepted by local entrepreneurs, ecotourists, and community.

6.3.5 Paraíba Aquarium Project

The Paraíba Aquarium (PA) was opened to the public in 2016. It relies on partnerships with universities, public agencies, nongovernmental organizations, and associations for the implementation of research, information exchange, and other activities related to marine conservation. Since its inauguration, the PA received 48,000 visitors, mainly families, surrounding community, educational institutions, researchers, scout groups, elderly groups, and disabled persons' groups. It develops environmental events and assisted programs (Fig. 6.4), through sensorial activities, lectures, dynamics, and environmental education and nature interpretation techniques, such as the actions "Unraveling the Marine creatures," "Young Investigator," and "Young handler." It promotes outdoor activities at the Seixas Beach Natural tide

Fig. 6.4 Students interacting with a shark tank. Photo by: Paraíba Aquarium

Pools by aims of a Snorkeling Marine Trail and a Coastal Walking Trail (through the interpretation of the coastal landscape at Seixas beach). It provides infrastructure, equipment, and technical information for graduate projects and undergraduate students.

6.3.6 IFSC Graduate Course in Marine Sciences Applied to Teaching

The graduate course in Marine Sciences Applied to Teaching has been offered since 2014 by the Federal Institute of Santa Catarina (IFSC) , at Itajaí campus. Its team is multidisciplinary, composed of fifteen teachers and administrative technicians in education from several areas of knowledge. The target public of the course is composed of teachers and managers of basic education from all educational areas (Fig. 6.5). The main objective of the course is to promote ocean literacy (see Chap. 1). The structuring axis of the course is to promote the implementation of integrative and interdisciplinary educational practices related to the marine environment in schools and scholar curriculum. The course comprises 400 h in which students develop knowledge related to marine biology, oceanography, meteorology, coastal management, environmental legislation, fundamentals of socioeconomic activities related to the marine environment such as fishing, aquaculture, and nautical principles, and disciplines related to pedagogical area.

Fig. 6.5 Field survey of the students of the graduate course in Marine Sciences Applied to Teaching of the Federal Institute of Santa Catarina (IFSC). Photo by: Benjamin Teixeira

6.4 Projects' Conceptual Characteristics

The conceptual characteristics of the projects were analyzed following the general guidelines stated by the National Program of Environmental Education-ProNEA (Brasil 2005), which adopts the Environmental Education model for Sustainable Societies and Global Responsibility (EESS). This approach was used in the works of Berchez et al. (2007, 2016), Pedrini et al. (2016), and Rhormens et al. (2017). According to these authors, the conceptual characteristics of EESS aim to change behaviors and emancipation from the individual to the whole community, being as follows: (a) transformative: facilitates changes in attitudes for the development of sustainable societies, (b) participative: encourages participation in collective mobilization, (c) embracing: involves all the various social groups, (d) permanent: performed as a continuous activity (or continued EE), (e) globalizing: acts directly on the reality of the activity and on achieving global dimensions, (f) ethical: respect for human beings and all life-forms, (g) interdisciplinary: integrates various forms of knowledge, (h) holistic: aims at transforming the individual as a whole, e.g., values and ethical concepts, (i) multiplying: aims at expanding activities through the formation of multipliers, (j) contextualized: considers the local social, geographical, and biological diversity in the projects (Table 6.2).

6.4.1 *The Target Public*

The analyzed projects target mainly on an audience composed of students of different levels, children and teenagers in the majority, but also teachers, school managers, tourists, and residents of the region where the project is carried out.

6.4.2 *The staff*

The composition of the teams of almost all the projects analyzed was not multidisciplinary, except for the Graduate Course in Marine Sciences Applied to Teaching of the IFSC due to the expertise and diversity of the specialists involved. This assumption is one of the most fundamental of EE, since each socioenvironmental issue to be addressed demands a specific multidisciplinary team. Berchez et al. (2007, 2016) and Pedrini and De-Paula (2011) point out that to obtain interdisciplinary a multidisciplinary team would be required.

Table 6.2 General results of the analysis of the conceptual characteristics of Environmental Education for Sustainable Societies in the selected projects

Conceptual characteristics	General results
Transforming	The actions implemented, in general, aim to enlighten conscious citizens about their local reality and worldview. Their goal is to promote a change of individual and collective attitudes, mainly related to tourism and its disconnection from their local reality as of the global ecosystem. All projects analyzed focus on the empowerment of coastal communities, the rescue of their traditional way of life. Specifically, the ProMar Project aims at the financial emancipation of the traditional people as microentrepreneurs
Embracing	The activities of most projects contemplate local social actors involved with a socioenvironmental problem, such as local community, students, teachers, businessmen, service providers, public officers, family, tourists, etc
Permanent	Although this is the most important and difficult feature to be put into practice, as it depends on funding, all the studied projects have shown continuous activities, some with more than 20 years of existence such as the Golfinho Rotador Project and Tamar, as they receive external sponsorship
Globalized	The analyzed projects address local problems and their connection with global threats such as disorientated tourism, and loss of biodiversity, among other problems in coastal areas, thereby promoting ocean literacy
Interdisciplinary	Interdisciplinary approach was present in all analyzed projects, but the financial means to hire a multidisciplinary team is still a problem. However, in some projects, there are professionals from several areas, as in the IFSC graduate course in Marine Sciences Applied to Teaching
Contextualized	All projects adopt an approach that contemplates the local social, biological, and geographic diversity in their pedagogical activities
Multiplying	Although all projects intend to train multiplier agents, some projects are dedicated to offering training courses to teachers from coastal schools. The involvement of all projects with local community schools allows the formation of multipliers of the activities. In addition, we can mention the training of ecotourism guides/conductors in the ProMar and Underwater Marine Trail projects
Holistic	All projects are structured around the axis of marine biodiversity, coastal and marine ecosystems, and their conservation with an integrative focus on changing of attitudes and values, thus avoiding an anthropocentric approach

6.4.3 Objectives

The objectives of the analyzed CMEE projects are related to their specific audiences, mainly tourists, and traditional communities, along the shore or in oceanic islands where they operate. Therefore, their goals vary. However, in all of them, CMEE could be identified as an instrument to cope with the rapid and negative socioenvironmental transformations. The actions aim to raise awareness, and educate, transform, and empower citizens and their community to mitigate the current socioenvironmental problems as well as to prevent future ones, ensuring sustainability of coastal and marine ecosystems and their social, geographical, and biological diversity.

6.4.4 The Methodological Approaches

Analyzing the methodological approach of the projects, it was possible to notice that they use different strategies, among which we can perceive two crucial points.

The first point is related to the appropriate formation and training of the team. At first, it's perceived in two different ways of running the whole processes related to the project (from elaboration, team formation, implementation, and maintenance): (a) dialogical bottom-up process, by aims of public opinion survey, meetings, and group deliberations, bringing together people from the region, as in the case of ProMar project; and (b) external model of project, in a top-down process, with little or no participation of local communities in decisions, gathering specific selected persons for the team, as in the case of the IFSC or Paraíba Aquarium.

The second point is related to the way of approaching the public. Great amounts of methodological strategies were found, from theoretical classes to practical activities in which ludic strategies are used in classrooms or outdoors. In short, methods can be based on indirect contact, such as the visualization of organisms in captivity or in the wild; or direct contact, as touching and cherishing marine organisms, mainly invertebrates, such as sea stars, and rays in touch tanks, such as also charismatic species like dolphins or sea turtles in captivity or in the wild.

We highlight the methods based in the direct exposure of the public to the ecosystem through underwater or terrestrial trails mediated by guides or monitors. By this method, the public is literally immersed in the visited environment visualizing organisms and the natural features of reefs, mangroves, beaches, and rocky shores. Among methodological strategies adopted, related to the interpersonal relationships, we must mention briefings, informal group conversation, interviews, questionnaires, video projections, lectures, theatrical plays, games, and simulations, all of them accompanied by debates or not. Unfortunately, we should remark the fact that many of these strategies are poorly adopted despite their high potential to sensitize and bring awareness to the target public.

It was noticed that the methodologies favor the ludic contact with the marine and coastal environment, which is in accordance with the assumptions of Berchez et al. (2007), Ghilardi and Berchez (2010), Pedrini and Saito (2014) and Pedrini et al. (2011, 2014, 2016). The objectives of these actions are focused on the construction of values, transformation of attitudes, and social emancipation, according to Quintas (2004), Berchez et al. (2016), and Rhormens et al. (2017), as seen in the premises of the Treaty on Environmental Education for Sustainable Societies and Global Responsibility (TEESS) (see more about TEESS in Chap. 2).

6.4.5 Efficacy Evaluation

Evaluative processes are eminently qualitative. They vary widely, according to the appraiser. Therefore, it is difficult to scale and compare projects. Only the Tamarzinhos Project or those projects which are linked to universities, such as

ProMar and Underwater Marine Trail Project, use quantitative assessments, enabling comparisons. All projects use some type of qualitative evaluation instrument or technique, such as questionnaires, interviews, personal reports, observation, recording behavioral displays, analysis of drawings, or informal group conversation (see more on evaluation in Chap. 5).

6.4.6 *Pedagogical Products*

As expected, the pedagogical activities vary greatly according to the available budget. The ones performed by small-scale projects generally require the acquisition and production of low-cost materials such as photo cards, e-book with simple presentations, posters, handmade games, black and white leaflets, simple field guides, as well as the promotion of local events. On the other hand, the projects with external financing have more expensive and sophisticated products such as elaborated field manuals, printed books, dynamic e-books, documentaries, plot movies performed by professionals, puppet theater, video games, clothes, bags, caps, various kinds of souvenirs, and customized materials and they also promote events of wide scope (see more on instructional materials in Chap. 8). However, greater effectiveness of activities cannot always be attributed to greater availability of material resources, as noticed for ProMar project and other community-based projects (Rhormens et al. 2017).

6.5 Difficulties

The main difficulty is the medium- and long-term financial sustainability of the projects, besides the maintenance of infrastructure and the impossibility of expansion to meet the growing demands of the public and the projects itself. The lack of governmental support is also a hindrance.

All analyzed projects count or have already counted on the support of some entity or institution to which they are linked, although financing is always problematic for the projects developed by public universities. Simple or community-based projects, such as Underwater Marine Trail and ProMar, have limited funds, but they thrive in meeting their goals. On the other hand, projects motivated by charismatic/flag species receive large sponsorships, as in the case of Tamar and the Golfinho Rotador, producing abundant publicity material and attending a target public of greater income. In fact, as Berchez et al. (2007, 2016), Pedrini et al. (2016), and Rhormens et al. (2017) stated, one of the important assumptions of transformative and emancipatory EE is its financial sustainability.

Another difficulty not linked to the sustainability of the projects, but of much importance, is the difficulty in achieving longer contact time and subsequent follow-up of the target public as seen in some projects, making it difficult to assess the efficacy of the activities. This occurs mainly in the case of tourists in marine or coastal protected areas, which make up the target public of Golfinho Rotador, ProMar, Tamar, and Underwater Marine Trail projects. In these projects, there is brief contact with the target public in short-term pedagogical activities, which makes it difficult to evaluate pedagogical effectiveness through pre-/posttests.

6.6 Solutions

The best way to ensure the sustainability of projects has been seeking financial support in medium- and long-term sponsorship programs at private or public corporations. This is the case of the Brazilian oil company, which has been financing the Tamar and Golfinho Rotador projects for decades. In the case of other projects, the financing is obtained from the federal and regional agencies and foundations for the support of researches, many times in the form of scholarships and daily or subsistence allowances. In the case of the graduate course in Marine Sciences Applied to Teaching of the IFSC, there is full funding obtained from the federal government and in the case of Paraíba Aquarium the financing comes in the form of the income obtained with the collection of tickets.

Obviously, the main consequence of the abovementioned financial support will be the training of qualified personnel, the acquisition of necessary resources and the possibility of producing efficient pedagogical materials and practices for the modification and consolidation of concepts about the environment by the target audiences of the projects.

In relation to the evaluation of the public, for now, only the projects leading with traditional communities are based on long-term activities and continuous contact, making it possible to evaluate the expected results, such as changes in attitudes. It is necessary to plan a strategy of long-term evaluation for the others that deal with different publics.

6.7 Conclusions

The projects analyzed in the present chapter can be considered well succeeded, but several projects like them face many difficulties to persist existing. We consider that a broad registration of all CMEE projects in Brazil is essential, considering not only their characterization but also their related bibliographic production, aiming the production of essential knowledge for the future development of CMEE in Brazil.

To make the CMEE's efforts more effective in Brazil, a greater articulation between the projects is needed, bringing integration and visibility to the CMEE, aiming its insertion in elementary and high school educational systems. In Brazil, there is still no guideline document with an exclusive focus on issues related to marine and coastal environments pedagogy. In this sense, it is urgent to rethink the basic educational curriculum in the light of marine sciences and CMEE aiming to promote effective ocean literacy.

Acknowledgments We would like to thank the colleagues Dr. José Martins Junior (Oceanographer of ICMBio and coordinator of the Rotating Dolphin Project); MSc. Karina Massei (Biologist Master of Sciences) of Paraíba Aquarium and MSc. Marta Smith Rhormens (Oceanographer and Master of Sciences) of the ProMar Project, for providing the essential data necessary for this document.

References

Aquário Paraíba (2018). https://www.aquarioparaiba.com.br/. Accessed 15 Sept 2018

Berchez FAS, Carvalhal F, Robim MJ (2005) Underwater interpretative trail: guidance to improve education and decrease ecological damage. Int J Environ Sustain Dev 4(2):128–139. https://doi.org/10.1504/IJESD.2005.007235

Berchez FAS, Ghilardi N, Robim MJ et al (2007) Projeto Trilha Subaquática: sugestão de diretrizes para a criação de modelos de educação ambiental em unidades de conservação ligadas a ecossistemas marinhos. OLAM Ciênc Tecnol 7(3):181–209

Berchez FAS, Ghilardi-Lopes NP, Correia MD et al (2016) Marine and coastal environmental education in the context of global climate changes – synthesis and subsidies for ReBentos (Coastal Benthic Habitats Monitoring Network). Braz J Oceanogr 64(spe2):137–156. https://doi.org/10.1590/S1679-875920160932064sp2

Brasil (2005) Programa Nacional de Educação Ambiental, 3rd ed. Ministério do Meio Ambiente e Ministério da Educação, Brasília

Coordenação-Geral de Educação Ambiental (COEA) (2001) Conclusões de um diagnóstico preliminar de projetos de Educação Ambiental nas escolas. In: MEC (Org) Panorama da Educação Ambiental no Ensino Fundamental. Secretaria de Educação Fundamental, Brasília, pp 93–100

EArte (2018) Estado da Arte da Pesquisa em Educação Ambiental no Brasil. http://www.earte.net/. Accessed 15 Sept 2018

Ghilardi NP, Berchez F (2010) Projeto Trilha Subaquática – modelos de educação ambiental marinha para o grande público. In: Pedrini AG (Org) Educação Ambiental Marinha e Costeira no Brasil. Eduerj, Rio de Janeiro, pp 71–92

González-Gaudiano E, Lorenzetti L (2009) Investigação em educação ambiental na América Latina: mapeando tendências. Educ Rev 25:191–211. http://doi.org/10.1590/S0102-46982009000300010

Hadel VF (2010) Programas de Visitas ao Centro de Biologia Marinha (USP): o monitor na mediação entre academia e o grande público. In: Pedrini AG (Org) Educação Ambiental Marinha e Costeira no Brasil. Eduerj, Rio de Janeiro, pp 93–114

Meadows DH (1989) Harvesting one hundredfold: key concepts and case studies in environmental education. United Nations Environment Programme, Nairobi

Pedrini AG (2010) Educação Ambiental Marinha e Costeira no Brasil; aportes para uma síntese. In: Pedrini AG (Org) Educação Ambiental Marinha e Costeira no Brasil. Eduerj, Rio de Janeiro, pp 19–31

Pedrini AG, De-Paula JC (2011) Educação Ambiental: críticas e propostas. In: Pedrini AG (Org) Educação Ambiental; reflexões e práticas contemporâneas. Vozes, Petrópolis, pp 90–146

Pedrini AG, Saito CH (Org) (2014) Paradigmas Metodológicos em Educação Ambiental. Vozes, Petrópolis

Pedrini AG, Maneschy FSA, Silva VG et al (2010) Projeto EduMar – Educação/Interpretação Ambiental Marinha para o mergulho recreativo. In: Pedrini AG (Org) Educação Ambiental Marinha e Costeira no Brasil. Eduerj, Rio de Janeiro, pp 143–165

Pedrini AG, Brotto DS, Messas T (2011) Gestão de Áreas Protegidas com Educação Ambiental Emancipatória pelo Ecoturismo Marinho: a proposta do Projeto EcoTurisMar. OLAM Ciênc Tecnol 3:6–81

Pedrini AG, Ursi S, Berchez F et al (2014) Metodologias em Educação Ambiental para a conservação socioambiental dos ecossistemas marinhos. In: Pedrini AG, Saito CH (Org) Paradigmas Metodológicos em Educação Ambiental. Vozes, Petrópolis, pp 132–151

Pedrini AG, Brotto DS, Pimentel DS et al (2016) Transformative and Emancipatory Environmental Education by Marine Ecotourism in the marine environmental protection area of Armação dos Búzios, Rio de Janeiro, Brazil. Rev Bras Ecot 9(3):445–470

Pegas F, Coghlan A, Rocha V (2012) An exploration of a mini-guide program: training local children in sea turtle conservation and ecotourism in Brazil. J Ecotourism 11:48–55. https://doi.org/10.1080/14724049.2011.631710

Quintas JS (2004) Educação no processo de gestão ambiental: uma proposta de educação ambiental transformadora e emancipatória. In: Layrargues PP (Coord) Identidades da Educação Ambiental Brasileira. Ministério do Meio Ambiente, Brasília, pp 113–140

Reigota M (2007) O estado da arte da pesquisa em Educação Ambiental no Brasil. Pesqui Educ Ambient 2(1):33–66

Rhormens MS, Pedrini AG, Ghilardi-Lopes N (2017) Implementation feasibility of a marine ecotourism product on the reef environments of the marine protected areas of Tinharé and Boipeba Islands (Cairu, Bahia, Brazil). Ocean Coast Manag 139:1–11. https://doi.org/10.1016/j.ocecoaman.2017.01.022

Silva Junior JM, Gerling C, Venturi E et al (2010) Férias Ecológicas: um programa de educação ambiental marinha em Fernando de Noronha. In: Pedrini AG (Org) Educação Ambiental Marinha e Costeira no Brasil. Eduerj, Rio de Janeiro, pp 201–226

Silva VRF, Mitraud SF, Ferraz MLCP et al (2015) Adaptive threat management framework: integrating people and turtles. Environ Dev Sustain 18(6):1541–1558. https://doi.org/10.1007/s10668-015-9716-0

Chapter 7
Opportunities and Challenges to Implement Environmental Education in Brazilian Coastal and Marine Protected Areas

Douglas de Souza Pimentel, Alexandre de Gusmão Pedrini, Larissa de Araújo Kawabe, Maria de Jesus Robim, Flavio Augusto de Souza Berchez, and Camila Pinto Meireles

Abstract The 1988 Brazilian Federal Constitution established the promotion of environmental education (EE) as a Government's public policy, which constitutes an important legal frame addressing this subject in Brazil, also considering the EE activities in coastal and marine protected areas (CMPAs). This chapter presents the

---

D. S. Pimentel (✉)
Science Department, Rio de Janeiro State University (UERJ), São Gonçalo, Rio de Janeiro, Brazil

Geography Department, Fluminense Federal University (UFF), Niterói, São Gonçalo, Rio de Janeiro, Brazil
e-mail: douglaspimentel@id.uff.br

A. G. Pedrini
Vegetal Biology Department, Rio de Janeiro State University (UERJ), Maracanã, Rio de Janeiro, Brazil
e-mail: pedrini@uerj.br

L. A. Kawabe
Center for Natural and Human Sciences, Federal University of ABC (UFABC), São Bernardo do Campo, São Paulo, Brazil

M. J. Robim
Forestry Institute of São Paulo (IF), State Environmental Secretary (SMA), São Paulo, São Paulo, Brazil

F. A. S. Berchez
Biosciences Institute and CienTec/USP, University of São Paulo (USP), São Paulo, São Paulo, Brazil

C. P. Meireles
Marine Biology and Coastal Environments Postgraduate Program, Fluminense Federal University (UFF), Niteroi, Rio de Janeiro, Brazil

Interdisciplinary Environmental Study Group (GEIA), Rio de Janeiro State University (UERJ), Rio de Janeiro, Brazil

N. P. Ghilardi-Lopes, F. A. S. Berchez (eds.), *Coastal and Marine Environmental Education*, Brazilian Marine Biodiversity,
https://doi.org/10.1007/978-3-030-05138-9_7

legal frame, concepts, and potentialities of EE in CMPAs. Also, it highlights some experiences of EE in Brazilian PAs particularly to identify gaps, potentialities, and specificities for the development of coastal and marine environmental education (CMEE). Despite the conceptual and legal support for the development of EE activities in CMPAs, managers have a great difficulty to achieve these goals given the lack of funding, resources, personnel, and training, among other challenges that will be presented throughout this chapter. Thereby, environmental interpretation strategies assume great importance as well as citizen science initiatives and partnerships between public entities, civil society, and educational institutions. Considering this, it is of fundamental importance to encourage the social participation and diversity of partnerships. Also, it is necessary to test and improve CMEE methodologies in order to potentiate the teaching and learning process and to strength the democratic participation in the CMPAs management. In addition, it is clear that the CMEE needs financial independence, which could be supplied by the ecotourism in the CMPAs, but such initiatives are still incipient in Brazil.

Keywords Treaty on Environmental Education for Sustainable Societies and Global Responsibility · Coastal and marine environmental education · Citizen science · Environmental interpretation research experiences · Environmental laws

7.1 Environmental Education in Coastal and Marine Protected Areas

The coastal and marine environmental education (CMEE) in Brazil began with the marine biologists' claims about the destruction of coastal and marine ecosystems. Its politicization grew gradually and was translated into actions more committed to the reality of local communities. After the establishment of environmental education (EE) as a government's public policy in Brazilian Federal Constitution in 1988 (Brasil 1988) and its reinforcement in the National Environmental Education Program (PRONEA), as well as the National Environmental Education Policy (PNEA—Federal Law n° 9.795/99—Brasil 1999), EE actions received a stronger legal and conceptual basis, as an educational process that should take place in formal and non-formal spaces. From 2004, the ProNEA (Brasil 2014) embodied the Treaty on Environmental Education for Sustainable Societies and Global Responsibility (TEESS) (see more on Chap. 2). The National System of Nature Protected Areas (SNUC, Federal Law n° 9.985/2000—Brasil 2000) and the National Strategy for Communication and Environmental Education in Protected Areas (ENCEA—Brasil 2007), published as part of the ProNEA, established a strong link between EE and protected areas (PAs).

Besides being spaces of environmental conservation, most of the Brazilian PAs categories were also conceived as spaces for education, part of a broader territory that encompasses traditional people, neighboring populations, diverse cultures and knowledge, identities, and diversities (Brasil 2000, 2016), giving plenty of opportunities for educators to work.

Thus, PAs offer a strategic opportunity for (re)approaching people to natural environments, promoting their social insertion by combining knowledge, reflections, challenges, affectivity, curiosity, imagination, and notion of belonging (Pimentel and Magro 2013; Pimentel et al. 2013; Toledo and Pelicioni 2005).

Pedrini (2010) analyzed different researches and actions related to CMEE in Brazil and defined some typologies as they were based on iconic species, ecosystems, social actors, simulacrums, and marine biology teaching (see more on Chap. 6). Here the inclusion of a sixth category related to the EE efforts to eliminate invasive alien species in the marine environment is proposed, as showed by the experiences of the "Sun Coral" Project (Projeto Coral Sol; Meireles et al. 2016).

Different PAs' visitation activities could be considered as a long-term investment in environmental conservation and a basic premise for the ecological tourism's role in protecting biodiversity. In the Brazilian institutional setting, EE is an integrating axis that could foster PAs' process of institutionalization and the consequent recognition of their importance to society (Pimentel and Magro 2012). Still, EE should be a main management concern to enable a closer relationship among managers, adjacent communities, and visitors (Pimentel and Magro 2013). Generally, to put EE in practice in these institutions, interpretative trails (IT) are developed, based on principles established for environmental interpretation (EI).

The environmental interpretation is based on visitors' practical experiences on natural areas. It is used to inform, raise awareness, educate and encourage the ability of observation and reflection about the relationships between people and nature (Tilden 1977). This EI conceptual frame could provide a structure for carrying out EE activities and contribute to the promotion of social mobilization in supporting the environmental conservation efforts and the social recognition of PAs (Pimentel and Magro 2012).

The EI should also promote citizens' scientific culture, acting as a communication tool of great interest for PAs managers because it could bring new values to foster a social involvement into the conservation efforts and a full citizenship's exercise. The interpretive messages create connections between the public and the meaning of the protected patrimony. They have a critical influence on visitors' perceptions and must be guided by ethics, honesty, and social responsibility (Brito 2008). Then, the EI is based on a set of principles and techniques to stimulate people's understanding of the environment through direct practical experience (Tilden 1977; Ham 1992), and therein lays the importance of EI to the PAs management.

Interpretive trails are often used as an EI management tool (Tabanez et al. 1997; Vasconcellos 2006). Thus, IT should stimulate new perceptions about nature based on a critical stance to allow this new environmental awareness and considering this, their guidelines show a great proximity to EE principles.

In view of the touristic branch of the SNUC, the EE activities are seen as a mean of sustainable territorial development, because it can also offer an economic alternative related to communities nearby to some no-take protected areas. Thus, Guimarães (2011) proposes that the protection of natural areas should be managed in an integrated, participatory, and cooperative manner, concomitantly with the development of environmental and patrimonial education programs, directed to all the related communities, their interests, and developed actions. However, many PAs face the challenge of lack of personnel and funding. The staff generally is involved not only with administrative tasks, but also with activities related to the managing and monitoring of the areas under their responsibility. Because of that, EE is usually relegated to the background, with occasional activities and on isolated days, usually following what we call the "environmental calendar," which has commemorative dates related to various environmental issues. In this context, the possibility of allying the interests of managers and coastal and marine scientists in citizen science projects to support the management of coastal and marine PAs (CMPAs), especially on those that allow all the different types of visitation (Ghilardi-Lopes 2015), could enable the data gathering and the monitoring at a broader spatial and temporal scales than which is currently possible with the available CMPAs' staff and scientists (Bonney et al. 2009; Ghilardi-Lopes 2015). In addition, it could lower the costs and allows citizens to act in the knowledge production, while promoting the awareness and sensitization of CMPAs visitors. In addition, the establishment of volunteer programs (as is already done, for example, by the Chico Mendes Institute of Biodiversity—ICMBio—in Brazil and by the Forest Foundation—FF—in the State of São Paulo), as well as partnerships with governmental and nongovernmental institutions can be a very useful possibility for the development of research and evaluation activities in Brazilian CMPAs.

The conflicts normally faced by EE in Brazilian CMPAs are related to the reconciling of the demands of resources by the social actors, as they are negatively impacting (Pedrini et al. 2007). Other impacts refer to the occurrence of intentional or unintentional trampling, organisms' withdrawal, and sediments re-suspension by tourists and local bathers (Santos et al. 2015). In addition, entrepreneurs offering "tourism products" in PAs without socio-environmental concern aggravate the problems. Considering this, the managers establish a series of restrictions that end up limiting their own actions. Thus it is urgently necessary to identify the environmental perceptions, uses, and presumptions of use of the coastal and marine resources as well as to construct and plan practical management actions. The ProMar (Rhormens et al. 2017—see Chap. 6) and Ecoturismar (Pedrini et al. 2011, 2016) are successfully tested models that aim to transform harmful practices into sustainable ones by transformative and emancipatory environmental education, which is possible through marine ecotourism (ME) in PAs.

The present chapter considers that, despite the EE conceptual advances in Brazil, the analysis of reported experiences can contribute to the implementation and improvement of CMEE activities and their management in Brazilian PAs, especially the specific actions required in coastal and marine ecosystems. Thus, some experiences of EE in PAs will be highlighted, particularly to identify gaps, potentialities, and specificities for the development of CMEE.

7.2 Brazilian CMPAs: Potentialities as EE Locus

The Brazilian coastal and marine zones are those located in the area between the coastline and the boundary of ocean-influenced environments up to the territorial sea limit (12 nautical miles—Brasil 1990), in addition to the exclusive economic zone (encompassing 200 miles off the coast—Brasil 2013).This area includes oceanic or coastal islands and the PAs, which specific conservation objective is the protection of coastal and marine habitats, resources, and culture (MMA 2002).

Brazilian coast occurs along a wide range of latitudinal variation, encompassing a high diversity of habitats, related to eight different ecoregions (Spalding et al. 2007), four geological and political regions (North, Northeast, Southeast and South Regions), and three large marine ecosystems (Sherman 1991) (Fig. 7.1). The strong differences between ecosystems, environmental and social conditions, importance, and sensitivities challenge the coastal and marine protection (Chatwin 2007) and environmental education programs.

Among them, the coral reefs, very abundant at the Eastern and Northeastern ecoregions, where they have a singular physiognomy, are among the most globally threatened ecosystems by bleaching and other consequences of climate change

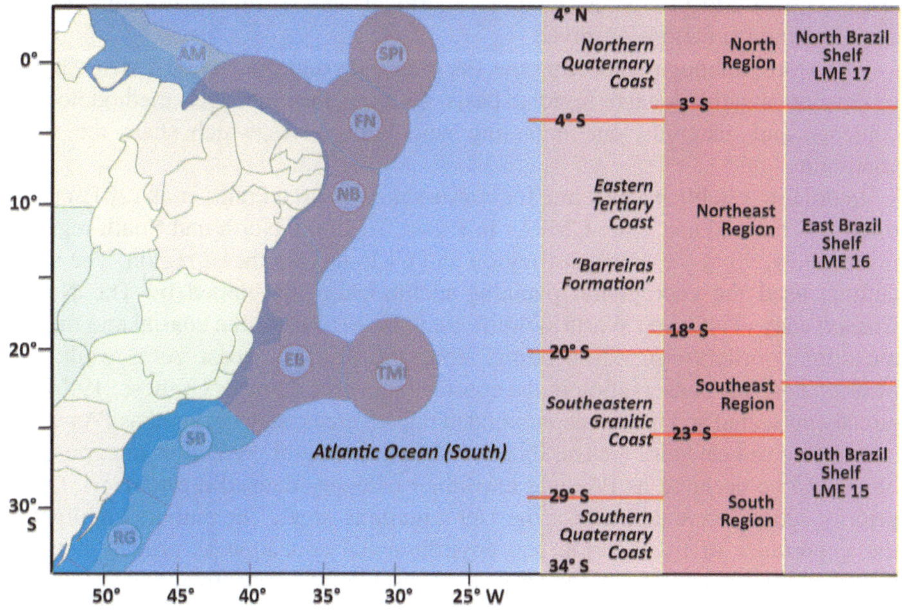

Fig. 7.1 Subdivisions of the Brazilian coast considering the eight *marine ecoregions of the world*: *AM* Amazonia, *SPI* São Pedro and São Paulo Islands, *FN* Fernando de Noronha and Atol das Rocas, *NB* Northeastern Brazil, *EB* Eastern Brazil, *TMI* Trindade and Martin Vaz Islands, *SB* Southeastern Brazil, *RG* Rio Grande, *geological divisions* (first column), *political divisions* (second column), and *large marine ecosystems*—LME (third column) (Spalding et al. 2007; Marone et al. 2010)

(Hughes et al. 2018; Leão et al. 2016). The Rhodolith beds, with similar distribution and occupying large extensions of the continental shelf, besides being the largest community of this kind worldwide and also severely endangered by ocean acidification (Amado-Filho et al. 2012; Horta et al. 2016), are almost unknown by the common people.

When comparing these ecosystems with, for instance, the rocky shores, related to crystalline rocks or the rocky substrate of sedimentary origin (sandstone reefs), occurring along all ecoregions (Coutinho et al. 2016), or the mangroves, which distribution extends from the Southeastern Brazil ecoregion up to the north (Schaeffer-Novelli et al. 2016), the differences of structure and function are considerable. Besides, strong variations in the structure of a single ecosystem could be found along the latitudinal gradient, such as in the case of estuaries, whose magnitude and amplitude of hydrological and tide regimes increases northward (Lana and Bernardino 2018), and mangroves, that shows distinct physiognomies, with higher vegetation sizes towards the north (Schaeffer-Novelli et al. 2016).

These differences, together with political regional characteristics, ecosystem services, economic utilization, ethic values, and other social aspects, also vary along the Brazilian coast (Zamboni and Nicolodi 2008). Thus, in the face of the fact of all these coastal habitats are endangered by multiple human impacts (Turra et al. 2013), the importance of environmental education is ratified to foster the knowledge about lesser known ecosystems as well as a critical stance on the different socio-environmental situations observed.

Therefore, although respecting the EE principles and directives, CMEE should be developed within local or regional basis, with specific objectives, methodologies, contents, and materials, encompassing neighboring PAs which share the same conditions.

In order to establish a baseline for conservation efforts, Prates et al. (2012) identified 208 Federal and State CMPAs in Brazil, 27 in the North and South regions, respectively, 90 in the Northeast region and 64 in the Southeast region. The same authors used the ecoregional planning methodology developed by The Nature Conservancy (TNC) and World Wildlife Fund (WWF) to define coastal and marine areas for priority conservation efforts along the Brazilian coast. As a result, 506 priority areas for conservation in the coastal zone were defined. Of those, 192 were already protected and it was recommended that 45 areas were declared as PAs in the sustainable use category (*sensu* SNUC), 42 in the integral protection category, and 58 should be declared as PAs and have their category defined *a posteriori*. Other priority actions were defined for the 169 remaining areas. The authors highlighted the importance of fiscalization and environmental education as priority actions, indicated for 324 and 286 of the proposed coastal areas, respectively.

The biodiversity conservation of the marine zone was considered very critical when what is already protected was compared with the total area defined as priority. Thus, the creation of 22 new marine PAs (7 in the sustainable use category; 1 in the integral protection category, and 14 with no category defined *a priori*) was recommended among the 102 proposed priority areas. For the areas in this zone, EE was not listed as the most relevant priority action, probably because the access to them is difficult for the general public. However, when considering the analysis for all priority actions in the proposed areas for conservation defined by the applied methodology, the document points EE as fundamental for 29 among them in the marine zone (Prates et al. 2012).

Considering the information presented by Prates et al. (2012), and all the uncertainties related to marine habitats, it can be hypothesized that coastal and marine environmental education actions are still very punctual and cover small areas of the Brazil's ecoregions, the existing PAs, and the new priority areas identified for conservation actions.

Citizen science (CS) is an interesting approach for CMEE conservation efforts in a context of lack of funding and personnel common in Brazilian PAs, by allying community volunteers with academic researchers (see more on Chaps. 4 and 8). Volunteers, with the help of professional researchers, can contribute in scientific research projects; especially in large temporal or spatial scale, which, at the same time, contributes to the scientific education of the participants (Bonney et al. 2009; Ghilardi-Lopes 2015). In Brazil, there are still a few CS projects, especially in marine and coastal environments (Thiel et al. 2014). The existing projects basically focus on the nektonic fauna as cetaceans, manatees and fish, such as the projects "Onde estão as baleias e os golfinhos?" (Where are the whales and the dolphins?—https://www.facebook.com/groups/baleiasgolfinhos.rj/members/), and "Mantas do Brasil" (Manta rays of Brazil—http://www.mantasdobrasil.org.br/), but the benthic community stays in the background.

Although citizen science is a recent field of study in the country, it shows great potential for the expansion of CMEE. The same could be said for partnerships development in the coastal and marine protected areas planning and management. However, these environmental education actions usually take place in easily accessible and more visited areas by general population, as rocky coastlines, sandy beaches, mangroves, and sandbanks. Nevertheless, some marine protected areas are difficult to access by the general public, thus CMEE actions should not focus on visitation in these cases, but considering the importance of the "know to preserve" theme, it is even more necessary to develop CMEE with a focus on little known ecosystems, such as deep coral reefs, oceanic islands, and even seagrass beds of accessible beaches that are not recognized by visitors and bathers. In this way, the challenge is to find strategies to bring these lesser known marine and coastal ecosystems to the population knowledge (Ghilardi-Lopes 2014).

7.3 EE Case Studies at Brazilian PAs

This section intends to present case studies on coastal and marine environmental education initiatives in PAs, illustrating the challenges for its implementation. In this sense, the potentialities, gaps, and specificities of the actions will be emphasized with the intention of contributing to future improvement of CMEE in PAs.

7.3.1 Anchieta Island State Park: PEIA

The Anchieta Island State Park (PEIA) is located in the municipality of Ubatuba, in the North Coast of São Paulo (SP) State. The Park has an area of 828 ha with around 22,600 m^2 of bathing areas, seven beaches with a total length of approximately 2 km, and five official land trails with a total extension of 3.6 km. The Park also has an underwater trail of 350 m and 2.4 ha of paths and green areas of public use (Fig. 7.2a). Currently the Park has an estimated visitation of 30,000 people per year. Most of the visitation takes place in the summer, and boats of companies registered in the PA perform the transportation of people from the continent to the island (Gallo et al. 2016).

Since 1992, the PEIA Public Use Program has offered environmental monitor training courses to teach concepts based on the constructivist methodology, in which the monitor learns about environment, procedures, and attitudes for the Park conservation, also developing skills to plan, reflect, and act in ecotourism and EE activities offered to visitors (Robim and Bussolotti 2003; Bussolotti and Robim 2005).

Fig. 7.2 (**a**) Panoramic view from Anchieta Island State Park to the continent (in the background). Photo by: Michael Swan, licensed in CC BY-ND 2.0; (**b**) Interpretative trail activity with rafts used by the Ecoturismar Project. Photo by: Daniel S. Brotto; (**c**) Coastal and marine environmental education activities at Itaipu Archeological Museum in Serra da Tiririca State Park. Photo by: Flávio S. Almeida (MAI collection); (**d**) Participants of the underwater trail of the "Environmental Education in Coastal and Marine Ecosystems" (Itaipu Marine Extractive Reserve). Photo by: Lara Guachalla

Considering the need to broaden educational frameworks and strategies for greater involvement with the school community, in 2003 the PEIA administration proposed the project "Environmental Education and Conservation of Water Resources of the Atlantic Forest" (São Paulo 2003). This was elaborated with the main objective of training teachers, environmental monitors, and students of the schools of Ubatuba for the water resources conservation. The strategies of action allowed the integration of the project participants for the planning and execution of the PAs terrestrial and marine environments interpretative trails. The results indicated that the methodology adopted led to the involvement of different partners (Educational State Board, Municipal Secretary of Education, Directors, Teachers, NGOs, Universities, and Environmental Educators of the region), favoring the expansion of educational actions, as well as the strengthening of the relationships between the protected area and the school community.

Among the partnership results, the underwater marine trail project, implemented in January 2002, aimed to develop and test models of CMEE activities (Berchez et al. 2005, 2007; Pedrini et al. 2008—see Chaps. 4, 6, and 8). This Project contributed significantly to the training of the teachers and monitors as well as the students' learning through schools' specific projects, under an environmental education perspective.

An intense emotional impact was observed in the evaluations, especially for those teachers who first experienced the diving activity. As for cognitive gains, abilities, and emotional involvement, they transcended the teaching of biology and ecology (Berchez et al. 2007, 2016). Among some characteristics that differentiate this CMEE initiative are the extent of 16 years, the development of nine distinct educational models (Ghilardi and Berchez 2010), and the extensive evaluation of results, resulting in several published papers (e.g., Berchez et al. 2005, 2007; Pedrini et al. 2010; Katon et al. 2014).

7.3.2 Marine Protected Area of Armação de Búzios

The Armação de Búzios Municipal act n° 135/2009 established the Marine Protected Area of Armação de Búzios (MPAAB) along the whole municipal seashore in Rio de Janeiro State. This act also established the Corals Municipal Natural Park in this area.

Most of tourists that visit Armação de Búzios come in travel ships that stay at the coast from 3 to 7 days. The Ecoturismar project considered that marine ecotourism (ME) would provide a feasible proposal for MPAAB management with the aid of transformative and emancipatory environmental education, as described by Pedrini et al. (2016). The methodology consisted in the identification of actors involved in the tourism chain as well as possible partnerships and support from local authorities and businessmen. The sample consisted of 85 persons present in the area during the summer season. The interpretive trail (IT) route on the rocky shore between the beaches of João Fernandes and João Fernandinho was selected. This IT focused on taxonomic characteristics of marine biodiversity, attractive biological species, and local ecology. Different didactic materials and procedures were developed, as four dialogical and interpretive lectures as well as interpretive cards and banners with photographs of local biodiversity and ecology. Support rafts with plastic bottles and other materials were developed (Fig. 7.2b). The evaluation of the marine trail as an ecotourism product was done by the confrontation of pre- and post-tests applied to tourists which participated in the activity. The tourists filled out a form with personal information, authorized the test and answered a pre-test. Then they attended a lecture about the local marine biota. At the end of the IT, they answered the post-test questionnaire. An increase in the holistic environmental perception and in the perception of socio-environmental interactions affecting the marine ecosystem was observed. The perception of negative impacts of conventional marine tourism on the environment and marine organisms also grew. The conclusion of the study is that TEEE could be developed as a community-based marine ecotourism (CBME) product, by means of underwater interpretative trails in protected Areas. The main challenges for permanent implementation of CBME in MPAAB were the disinterest of the majority of the local trade by sustainable marine ecotourism that generates less financial profit than the mass tourism, as well the degradation and death of the coral reef ecosystem that is the main attraction of marine ecotourism due to the inefficiency of state environmental management. The main benefits are the demand of tourists, mostly foreigners from temperate countries that come in search of sun and tropical beach tourism in Brazil; the curiosity and interest for the beauty of the marine ecosystem fauna and flora of the region, which is characterized by upwelling on summer and water temperatures of about 18 °C; the income generation for "caiçara" traditional community, and an increase in the collection of municipal taxes. Thus, in a broader social-environmental vision this form of marine ecotourism would bring enormous advantages to local communities.

7.3.3 Serra da Tiririca State Park Region

The Serra da Tiririca State Park (PESET) and the Itaipu Marine Extractive Reserve (RESEX Itaipu) are also located at Rio de Janeiro State. Besides the Atlantic Rainforest Biome, the two PAs encompass marine and coastal areas that include three islands, rocky coastlines, coves, sandy beaches, sambaqui dunes (archeological site), and a lagoon estuary. The Itaipu Archeological Museum (MAI) and the Professor Marcos Waldemar de Freitas Reis Municipal School established a partnership with researchers of the Interdisciplinary Environmental Study Group (GEIA) from Rio de Janeiro State University and Park managers to develop different interpretative trails for CMEE (see also Chap. 8). They considered important that the environment interpretation activities should be related to the formal educational spaces but act as touristic products as well. Thus, they could ratify the local institutions' relevance to enable the use of the regional ecosystems both as educational and touristic spaces (Fig. 7.2c, d).

This partnership was fundamental to the project development. The different academic works contributed to the inter-institutional cooperation due to the lack of resources and personnel for environmental conservation and management of the Park's public use. Also, the activities necessarily addressed socio-environmental issues ratifying the multidisciplinary basis of the process. It was also considered that the developed ITs were of enormous importance to empower local people for environmental management based on EE key precepts of critical learning.

The first motivations for the development of interpretative trails in the region were based on the lack of knowledge about the local socio-environmental characteristics. But the project demanded by MAI was limited to aspects related to biodiversity. The GEIA researches considered it an opportunity to develop teaching, research, and outreach activities with a broader look. However, there is the challenge of aligning the different institutional agendas. What is clear is that all won with the partnership and managed to meet, at least in part, the budget deficiencies for the realization of coastal and marine environmental education.

7.4 Final Considerations

Although there is a conceptual and a legal support for the development of environmental education activities in coastal and marine protected areas, managers probably have difficulties to achieve this goal considering the lack of funding, personnel, and training. Besides these limitations, coastal and marine environmental education is dependent on the specificities for the development of these activities, such as the visitation security issues, the need for basic equipment, and brief training for its use as well as by climatic conditions.

The CMPAs have the richest and most sensitive habitats of the Brazilian coast and represent privileged sites of great potential attraction for the visiting public and marine environmental education and interpretation programs. However, the knowledge of the coastal ecosystems is little explored by PAs public use program and by teaching activities in the coastal regions schools.

It is considered that visiting protected marine environments and associated activities may negatively impact species and/or their habitats. However, it is also known that the visitation in these places, accompanied by a structured CMEE program, with scientific bases, could reach significant results to connect people with these CMPAs and to mitigate negative impacts on the species and habitats. In fact, some transformative and emancipatory environmental education and citizen science initiatives had already been tested and pointed into this direction (Berchez et al. 2007, 2016; Pedrini et al. 2011; Lodi and Tardin 2018). So, the great challenge is to encourage the visitor to know and enjoy these environments in a responsible way, stimulating the appreciation and understanding of marine life in search for public support and local communities' participation in the environmental education actions. Also, there is a need to promote values of "low impact" visitation in Brazilian CMPAs, prioritizing the elaboration and implementation of public use plans, with continuous monitoring. All this movement could ultimately generate a greater public support for the environmental conservation.

However, it is still necessary to evaluate (see more on Chap. 5) and improve CMEE methodologies (see more on Chap. 4) to potentiate the teaching and learning process and to strengthen the democratic participation in the CMPAs management, effectively contributing to the exercise of citizenship. Thereby, it is clear that the coastal and marine environmental education needs for a financial independence that could be supplied by the ecotourism in the CMPAs. Thus, the desired socio-environmental benefits of EE should be fostered by the continuous development of the conceptual frame and new approaches, such as those promoted by public and private partnerships. Besides this, other challenges are imposed, such as: (1) difficulties in addressing little known coastal and marine ecosystems; (2) CMPAs where visitation is not allowed or viable; (3) interconnection among similar actions developed by several actors in different places, and (4) troubles in identifying indicators and developing monitoring strategies of the impacts of CMEE actions.

Thus, the coastal and marine environmental education in Brazilian PAs starts from a good legal frame and political structure that places the EE itself as the central axis. The works always involve the visitors (local community or tourists) with the aim of bringing knowledge, reflection and sensitization and critical posture to the CMPAs visitors. However the challenges still stand in the lack of resources to enable the different activities. It is also important to emphasize the need to improve the management and monitoring of such marine and coastal ecotourism initiatives. As a suggestion, the present chapter proposes the partnerships development among public institutions (museums, schools, and PAs), civil society, and other actors for the development of CMEE actions in the CMPAs. The roles of implementation of these activities, strategies, and monitoring would be shared between the parties, in a continuous reassessment process so that the CMEE can achieve its objectives of preparing critical citizens, engaged in the development of a truly democratic environmental management.

Acknowledgements We thank Peterson Lásaro Lopes for the elaboration of Fig. 7.1.

References

Amado-Filho GM, Moura RL, Bastos AC et al (2012) Rhodolith beds are major $CaCO_3$ bio-factories in the tropical South West Atlantic. PLoS One 7:1–6. https://doi.org/10.1371/journal.pone.0035171

Berchez F, Carvalhal F, Robim MJ (2005) Underwater interpretative trail - guidance to improve education and decrease ecological damage. Int J Environ Sustain Dev 4:128–139. https://doi.org/10.1504/IJESD.2005.007235

Berchez F, Ghilardi N, Robim MJ et al (2007) Projeto Trilha Subaquática: sugestão de diretrizes para a criação de modelos e educação ambiental em unidades de conservação ligadas aos ecossistemas marinhos. OLAM Cienc Tecnol 7:181–209

Berchez F, Ghilardi-Lopes NP, Correia MD et al (2016) Marine and coastal environmental education in the context of global climate changes - synthesis and subsidies for ReBentos (Coastal Benthic Habitats Monitoring Network). Braz J Oceanogr 64(spe2):137–156. http://doi.org/10.1590/S1679-875920160932064sp2

Bonney R, Cooper CB, Dickinson J et al (2009) Citizen science: a developing tool for expanding science knowledge and scientific literacy. Bioscience 59(11):977–984. https://doi.org/10.1525/bio.2009.59.11.9

Brasil (1988) Constituição da República Federativa do Brasil. Senado, Brasília

Brasil (1990) Decreto n° 99.165, de 12 de março de 1990. Promulga a Convenção das Nações Unidas sobre o Direito do Mar. Diário Oficial da União, Brasília

Brasil (1999) Lei nº 9.795, de 27 de abril de 1999. Dispõe sobre a educação ambiental, institui a política de educação ambiental e dá outras providências. Diário Oficial da União, Brasília

Brasil (2000) Lei nº 9.985, de 18 de julho de 2000. Regulamenta o art. 225, § 1o, incisos I, II, III e VII da Constituição Federal, institui o Sistema Nacional de Unidades de Conservação da Natureza e dá outras providências. Diário Oficial da União, Brasília

Brasil (2007) Diretrizes para Estratégia Nacional de Comunicação e Educação Ambiental em Unidades de Conservação. Ministério do Meio Ambiente e ICMBio, Brasília

Brasil (2013) Projeto de Lei n° 6.969 de 2013. Institui a Política Nacional para a Conservação e o Uso Sustentável do Bioma Marinho Brasileiro (PNCMar) e dá outras providências. Brasília

Brasil (2014) Programa Nacional de Educação Ambiental (ProNEA). 4ª ed. Ministério do Meio Ambiente, Brasília

Brasil (2016) Educação ambiental em unidades de conservação: ações voltadas para comunidades escolares no contexto da gestão pública da Biodiversidade. Instituto Chico Mendes de Conservação da Biodiversidade (ICMBio) e WWF-Brasil, Brasília

Brito LM (2008) O guia-intérprete: mediador intercultural. Rev Turismo Desenvolvimento 10:67–84

Bussolotti JM, Robim MJ (2005) Avaliação da capacitação de monitores ambientais do Parque Estadual da Ilha Anchieta. IF Série Registros 28:13–37

Chatwin A (2007) Priorities for coastal and marine conservation in South America. The Nature Conservancy, Virginia

Coutinho R, Yaginuma LE, Siviero F et al (2016) Studies on benthic communities of rocky shores on the Brazilian coast and climate change monitoring: status of knowledge and challenges. Braz J Oceanogr 64(spe2):27–36. http://doi.org/10.1590/S1679-875920161015064sp2

Gallo H Jr, Pedro ML, Moreira PS, Robim MJ (2016) Indicadores de gestão do uso público no Parque Estadual da Ilha Anchieta. OLAM Cienc Tecnol 16(1-2):82–109

Ghilardi NP, Berchez F (2010) Projeto Trilha Subaquática - modelos de educação ambiental marinha para o grande público. In: Pedrini AG (ed) Educação ambiental marinha e costeira no Brasil. EdUERJ, Rio de Janeiro, pp 71–91

Ghilardi-Lopes NP (2014) Engaging people in the conservation of 'non-charismatic' benthic marine organisms: a Brazilian case study. Eingana 37:11–13

Ghilardi-Lopes NP (2015) Citizen science combined with environmental education can be a powerful tool for coastal-marine management. J Coast Zone Manag 18:407. https://doi.org/10.4172/jczm.1000407

Guimarães STL (2011) Aspectos da percepção e valoração de paisagens do Núcleo Santa Virgínia, Parque Estadual da Serra do Mar, SP, Brasil. OLAM Cienc Tecnol 11(2):228–248

Ham SH (1992) Interpretación ambiental: una guia práctica para gente con grandes ideas y presupuestos pequeños. Fulcrum/North American Press, Golden Colorado

Horta PA, Riul P, Amado Filho GM et al (2016) Rhodoliths in Brazil: current knowledge and potential impacts of climate change. Braz J Oceanogr 64(spe2):117–136. http://doi.org/10.1590/S1679-875920160870064sp2

Hughes TP, Anderson KD, Connolly SR et al (2018) Spatial and temporal patterns of mass bleaching of corals in the Anthropocene. Science 359:80–83. https://doi.org/10.1126/science.aan8048

Katon GF, Berchez F, Oliveira VM, Ursi S (2014) The influence of field environmental education activities on teacher's marine environmental perception. Rev Educ Cienc 15:83–84

Lana PC, Bernardino AF (eds) (2018) Brazilian estuaries: a benthic perspective, vol 1. Brazilian marine biodiversity. Springer, Cham

Leão ZMAN, Kikuchi RKP, Ferreira BP et al (2016) Brazilian coral reefs in a period of global change: a synthesis. Braz J Oceanogr 64(spe2):97–116. http://doi.org/10.1590/S1679-875920160916064sp2

Lodi L, Tardin R (2018) Citizen science contributes to the understanding of the occurrence and distribution of cetaceans in southeastern Brazil – a case study. Ocean Coast. Manag 158:45–55. https://doi.org/10.1016/j.ocecoaman.2018.03.029

Marone E, Knoppers B, Souza WFL, Silveira IC, Godoi SS (2010) The Brazil current: physical–biogeochemical domains. In: Liu KK, Atkinson L, Quinones R, Talaue-McManus L (eds) Carbon and nutrient fluxes in continental margins. A global synthesis, Global change-the IGBP series. Springer, Berlin, pp 153–169

Meireles CP, Creed JC, Pimentel DS (2016) A Educação Ambiental no controle da bioinvasão marinha por Coral-Sol (*Tubastraea* spp. Anthozoa, Dendrophylliidae) em Angra dos Reis (RJ, Brasil). Ambient Sustentable 2:1–21. https://doi.org/10.17979/ams.2015.02.020.1612

Ministério do Meio Ambiente (2002) Avaliação e identificação de áreas e ações prioritárias para a conservação, utilização sustentável e repartição dos benefícios da biodiversidade nos biomas brasileiros. Ministério do Meio Ambiente, Brasília

Pedrini AG (2010) Educação Ambiental Marinha e Costeira no Brasil; aportes para uma síntese. In: Pedrini AG (Org) Educação Ambiental Marinha e Costeira no Brasil. Eduerj, Rio de Janeiro, pp 19–31

Pedrini AG, Costa C, Newton T et al. (2007) Efeitos ambientais da visitação turística em áreas protegidas marinhas: estudo de caso na Piscina Natural Marinha, Parque Estadual da Ilha Anchieta, Ubatuba, São Paulo, Brasil. OLAM Cienc Tecnol 7(1):678–696

Pedrini AG, Dutra D, Robim MJ, Martins SL (2008) Gestão das Áreas Protegidas e Avaliação de Educação Ambiental no Ecoturismo: Estudo de Caso com o Projeto Trilha Subaquática - Educação Ambiental nos Ecossistemas Marinhos no Parque Estadual da Ilha Anchieta, Ubatuba, São Paulo, Brasil. OLAM Cienc Tecnol 8:31–55

Pedrini AG, Messas TP, Pereira ES, Ghilardi-Lopes NP, Berchez FA (2010) Educação Ambiental pelo Ecoturismo numa trilha marinha no Parque Estadual da Ilha Anchieta, Ubatuba (SP). Rev Bras Ecoturismo 3:428–459

Pedrini AG, Brotto DS, Lopes MC, Messas TP (2011) Gestão de áreas protegidas com educação ambiental emancipatória pelo ecoturismo marinho: a proposta do projeto Ecoturismar. OLAM Cienc Tecnol 3:6–81

Pedrini AG, Brotto DS, Pimentel DS, Behrends E, Junqueira A (2016) Transformative and Emancipatory Environmental Education by Marine Ecotourism in the marine environmental protection area of Armação dos Búzios, Rio de Janeiro, Brazil. Rev Bras Ecoturismo 9(3):445–470

Pimentel DS, Magro TC (2012) Indicadores para a gestão da inserção social de parques. OLAM Cienc Tecnol 12(1-2):254–277

Pimentel DS, Magro TC (2013) Diferentes dimensões da Educação Ambiental para a inserção social de parques. Rev Bras Educ Ambient 7:44–50

Pimentel DS, Magro TC, Silva-Filho DF (2013) Imagens da Conservação: Em busca do apoio público para a gestão de Unidades de Conservação. Teoria Soc (UFMG) 19:144–168

Prates AP, Gonçalves MA, Rosa MR (2012) Panorama da conservação dos ecossistemas costeiros e marinhos no Brasil. Ministério do Meio Ambiente, Brasília

Rhormens MS, Pedrini AG, Ghilardi-Lopes NP (2017) Implementation feasibility of a marine ecotourism product on the reef environments of the marine protected area of Tinharé and Boipeba Islands (Cairu, Bahia, Brazil). Ocean Coast Manag 139:1–11. https://doi.org/10.1016/j.ocecoaman.2017.01.022

Robim MJ, Bussolotti JM (2003) Capacitação de monitores ambientais do Parque Estadual da Ilha Anchieta In: Bahl M (Org) Mercado Turístico: áreas de atuação. Roca, São Paulo, pp 213–220

Santos GS, Burgos DC, Lira SMA, Scwamborn R (2015) The impact of trampling on reef macrobenthos in northeastern Brazil: how effective are current conservation strategies? Environ Manage 56(4):847–858. https://doi.org/10.1007/s00267-015-0552-7

São Paulo (2003) Educação Ambiental: o que se pensa e o que se faz. Secretaria de Estado do Meio Ambiente, São Paulo

Schaeffer-Novelli Y, Soriano-Sierra EJ, Vale CC et al (2016) Climate changes in mangrove forests and salt marshes. Braz J Oceanogr 64(spe2):37–52. http://doi.org/10.1590/S1679-875920160919064sp2

Sherman K (1991) The large marine ecosystem concept: research and management strategy for living marine resources. Ecol Appl 1(4):349–360. https://doi.org/10.2307/1941896

Spalding MD, Fox HE, Allen GR et al (2007) Marine ecoregions of the world: a bioregionalization of coastal and shelf areas. Bioscience 57(7):573–583. https://doi.org/10.1641/B570707

Tabanez MF, Pádua SM, Souza MGI (1997) Avaliação de trilhas interpretativas para educação ambiental. In: Pádua S, Tabanez MF (Orgs) Educação ambiental: caminhos trilhados no Brasil. IPE, Nazaré Paulista, pp 89–101

Thiel M, Penna-Diaz MA, Luna-Jorquera G, Salas S, Sellanes J, Stotz W (2014) Citizen Scientists and Marine Research: Volunteer Participants, Their Contributions, and Projection for the Future. Oceanogr Mar Biol 52:257–314. https://doi.org/10.1201/B17143

Tilden F (1977) Interpreting our heritage, 3rd edn. University of North Carolina Press, Chapel Hill

Toledo RF, Pelicioni MC (2005) Educação ambiental em unidades de conservação. In: Phillippi Jr. A, Pelicioni MCF (Orgs) Educação Ambiental e Sustentabilidade. Manole, Barueri, pp 749–769

Turra A, Cróquer A, Carranza A et al. (2013) Global environmental changes: setting priorities for Latin American coastal habitats. Global Change Biology 19:1965–1969. https://doi.org/10.1111/gcb.12186

Vasconcellos JM (2006) Educação e interpretação ambiental em unidades de conservação. Curitiba 3(4):1–86

Zamboni A, Nicolodi JL (eds) (2008) Macrodiagnóstico da Zona Costeira e Marinha do Brasil. Ministério do Meio Ambiente, Secretaria de Mudanças Climáticas e Qualidade Ambiental, Brasília

Chapter 8
Didactic Materials as Resources for the Promotion of Coastal and Marine Environmental Education

Natalia Pirani Ghilardi-Lopes, Douglas de Souza Pimentel, Laura Pioli Kremer, Renato de Almeida, and Camila Pinto Meireles

Abstract Creativity and playfulness are important skills that educators use to promote environmental awareness and changes in beliefs, attitudes, and values. The production of reusable and easily available didactic materials can assist in this process. The purpose of this chapter is to present some teaching and learning didactic materials developed in Brazil for different marine and coastal environmental education activities. We will present some examples of materials produced for interpretive trails, as well as books, guides, folders, radio programs, games, and materials for

N. P. Ghilardi-Lopes (✉)
Center for Natural and Human Sciences, Federal University of ABC (UFABC), São Bernardo do Campo, São Paulo, Brazil
e-mail: natalia.lopes@ufabc.edu.br

D. S. Pimentel
Science Department, Rio de Janeiro State University (UERJ), São Gonçalo, Rio de Janeiro, Brazil

Geography Department, Fluminense Federal University (UFF), Niterói, São Gonçalo, Rio de Janeiro, Brazil
e-mail: douglaspimentel@id.uff.br

L. P. Kremer
Campus Itajaí, Instituto Federal de Santa Catarina (IFSC), Itajaí, Santa Catarina, Brazil
e-mail: laura.kremer@ifsc.edu.br

R. Almeida
Centro de Ciências Agrárias, Ambientais e Biológicas, Universidade Federal do Recôncavo da Bahia (UFRB), Cruz das Almas, Bahia, Brazil
e-mail: renato.almeida@ufrb.edu.br

C. P. Meireles
Marine Biology and Coastal Environments Postgraduate Program, Fluminense Federal University (UFF), Niteroi, Rio de Janeiro, Brazil

Interdisciplinary Environmental Study Group (GEIA), Rio de Janeiro State University (UERJ), São Gonçalo, Rio de Janeiro, Brazil
e-mail: camilameireles@id.uff.br

N. P. Ghilardi-Lopes, F. A. S. Berchez (eds.), *Coastal and Marine Environmental Education*, Brazilian Marine Biodiversity,
https://doi.org/10.1007/978-3-030-05138-9_8

exhibitions. Also, we will reflect upon the necessity of enhancing the dissemination and sharing of these materials among Brazilian environmental educators as well as their proper evaluation.

Keywords Didactic materials · Support materials · Environmental education strategies

8.1 The Importance of Didactic Materials for Environmental Education Activities

Instructional materials are resources designed to support or supplement instruction (Remillard and Heck 2014), which can have a substantial effect on learning. There are two types of didactic/instructional materials according to the actors of the educational process: teaching materials, which provide educators with resources to guide and support the learning process of students; and learning materials, which support learners during the process of learning, such as books, games, worksheets, etc. (Padrón et al. 2005). In both cases, it is advantageous if the material can be used under different contexts (reusability) and easily available for users through, for example, online repositories (Koppi et al. 2005).

In South America, especially in Brazil, there are few repositories in which to find environmental education (EE) instructional materials, such as the International Bank of Educational Objects linked to the Brazilian Ministry of Education (Afonso et al. 2011), the Digital School (http://www.escoladigital.org.br/), or the sites related to sectors of Brazilian government related to education and environment (e.g., Environmental Ministry—http://www.mma.gov.br/, Chico Mendes Institute for Biodiversity Conservation—http://www.icmbio.gov.br/portal/, and Education Ministry—http://mec.gov.br). Even when the materials are found, many times the limitation of funds induces educators to create their own didactic materials to support EE activities. This can be positive at one hand, since the produced materials are generally inserted in those teachers and students' context, exploring their creativity and even using low-cost elements for their elaboration. Also, the process of planning and developing those materials can bring itself an educational gain when learners and educators work collaboratively (Gokhale 1995). On the other hand, this can be detrimental to the expected educational goals of proposed activities, since the materials (or learning objects) available at digital repositories are generally evaluated in educational terms, so they can be considered reliable based on previous experiences (Chiappe Laverde et al. 2007) and present the characteristics of reusability, generativity, adaptability, and scalability (Wiley 2002).In synthesis, it is important to promote the creation of didactic materials, but also their sharing and continuous evaluation is fundamental (see more on evaluation in Chap. 5).

In the next sections, we will present some didactic materials developed for Brazilian EE activities, especially materials produced for interpretive trails, as well as books, guides, folders, radio programs, games, and materials for exhibitions.

8.2 Materials for Interpretive Trails (ITs)

Interpretive trails (ITs) are structured to provide environmental interpretation (EI) directly through the contact between individuals and the natural environment. They can approach a central theme and use interpretive points usually mediated by playful activities. As an example of a specific material for the implementation of marine and coastal ITs, we can mention the use of traffic cones as underwater magnifiers (Fig. 8.1) by coupling a glass or acrylic plate to their basal portion. Due to their simplicity and low cost, they have been used in ITs in the states of São Paulo (SP) (Ghilardi and Berchez 2010), Rio de Janeiro (RJ) (Meireles et al. 2013), and Bahia (BA) (Rhormens et al. 2017). In RJ, another material was recently tested for the "Itaipu Waters Circuit" by the Interdisciplinary Environmental Study Group (GEIA), which is a floating polypropylene rope with pending cards containing information about the EI stopping points along the trail route. The cards can be made from the reuse of plastic packaging (Fig. 8.1). In this way, visitors follow the rope with one hand and find the cards, which present questions and activities on a chosen theme or the identification of organisms, aiming at the use of the senses and a greater perception of the marine environment. Coastal and marine environmental education (CMEE) activities need to cause minimum impact and for such a purpose, their coupling with briefings, prohibition of flippers (except for monitors), and using of buoyancy floats as support is indicated. In Anchieta Island State Park (SP), the Underwater Marine Trail Project (Berchez et al. 2007) uses floating rafts made of PET bottles to lead the participants along the way. The same support material was used at Armação de Búzios (RJ) by the Ecoturismar project (Pedrini et al. 2011) (see more on these two last projects in Chaps. 6 and 7).

Another point to be considered in the development of ITs is the replication of the activity by different educators. Both in São Paulo and Rio de Janeiro States, the need to create a standardized protocol for the underwater interpretive trails was verified, so that all environmental educators could approach the IT theme in the same

Fig. 8.1 (**a**) Traffic cones adapted for the use as an underwater magnifier by the participants of the Ecorais Project (Búzios, Rio de Janeiro State) and (**b**) floating rope with plastic cards used in the "Itaipu Waters Circuit" IT (GEIA-UERJ at Niterói, Rio de Janeiro State). Photos by: Camila P. Meireles and Lara Guachalla

way. At the Sun Coral Project (Projeto Coral-Sol) IT in Angra dos Reis (RJ), the team had to create a "Guide to the Guides," so that the educators could properly develop the contents considered essential for the trail (Meireles et al. 2016).

Finally, it is important to consider that for a democratic CMEE process, the inclusion of participants who present any kind of physical disability is fundamental. Nowadays, adapted equipment for wheelchair users is already available to the practice of terrestrial and underwater trails (Watson 1982; Giraldo and D'amico 2016). Also, interpretive points that sensory exploit the environment for visually impaired people has long been used (Seven 1980; Karyn Benton 2011). However, these strategies need to be better exploited in Brazil for the marine environment.

8.3 Books, Guides, and Folders

Books and guides can be very helpful materials for the use with EE activities developed for specific contexts or with activities, which can happen in nonspecific places and/or contexts.

For instance, we can mention several materials which were developed for EE activities on Serra da Tiririca region (RJ) (see also Chap. 7 for more information on this project). Four main institutions were involved in this process: (1) the Serra da Tiririca State Park (PESET) administration (Pimentel et al. 2004); (2) the Archaeological Museum of Itaipu (MAI), which aims at promoting the community's recognition of the environmental and historical heritage; (3) the local schools, particularly the Professor Marcos Waldemar de Freitas Reis Municipal School, which look for incorporating EE in its curriculum; and (4) the Rio de Janeiro State University (UERJ), which search for ways to promote activities of teaching, academic research, and outreach that are important for the accomplishment of EE in formal and nonformal spaces of teaching.

A project entitled "Monitoring and Diagnosis of Itaipu Lagoon Health and Fisheries Resources" was proposed, encompassing interdisciplinary actions in the three academic axes (research, teaching, and outreach) in order to promote interactions among the partner institutions (Barcellos et al. 2015; Fernandes et al. 2011). The workshops promoted during the project stimulated the discussion on socio-environmental issues and counted with the participation of local social actors. The planned EE activities involved field trips, lectures, practical observations, fish biometric measures, and fish identification based on local fishermen knowledge, always involving the school's teachers to address regional environmental issues in a broader way. Important pedagogical materials were developed as EE activities indicated some needs and goals. The first material was the "Morro das Andorinhas Interpretive Trail Guide" (Barcellos et al. 2015), developed by an UERJ undergraduate student considering the environmental interpretation (EI) as an enabling tool for EE that could bring positive impacts related to public use in parks and an interdisciplinary learning process at school. The IT aims to foster new perceptions about nature to stimulate a critical stance by students and has six interpretative points, which considered several dimensions of information (biological, ecological, socio-environmental, legal, and conservational). The guide presents a map of the place, contributing to geographical location exercises. There is information about how to proceed on the trail. All the six interpretive points have designations and important information, some questions, and discussion. The 5th and 6th points focus on coastal and marine ecosystems and human impacts. In the end, the students are stimulated to register perceptions by writing or drawing. A practical result was the use of this IT guide as a complement to the school curriculum, as indicated by the Brazilian National Curriculum Parameters (PCNs) (Brasil 1998). Nowadays, such activities are part of the Professor Marcos Waldemar de Freitas Reis Municipal School pedagogical program, and other ITs and guides were produced (Pimentel et al. 2017).

The second pedagogical product resulting from this project was a para-didactic book (Pimentel et al. 2013), developed by UERJ professors and students with the participation of MAI staff as well as teachers and students from Professor Marcos Waldemar de Freitas Reis Municipal School. A history was created, and the text was printed to papers with some blank space to be filled by the readers (children) with drawings of the situations presented. This activity was oriented by a school teacher and the illustrations were incorporated into the book (Fig. 8.2).

The characters imagined and drawn by the children were used to provide information about Serra da Tiririca features, its prehistory and history, developmental characteristics of the neighborhoods and surrounding cities, about the PESET and MAI as well as local biomes, including their organisms, from benthic invertebrates to vertebrates.

This project and its products supported an institutional integration as well as the recognition by the local community of its own relevance. The didactic material

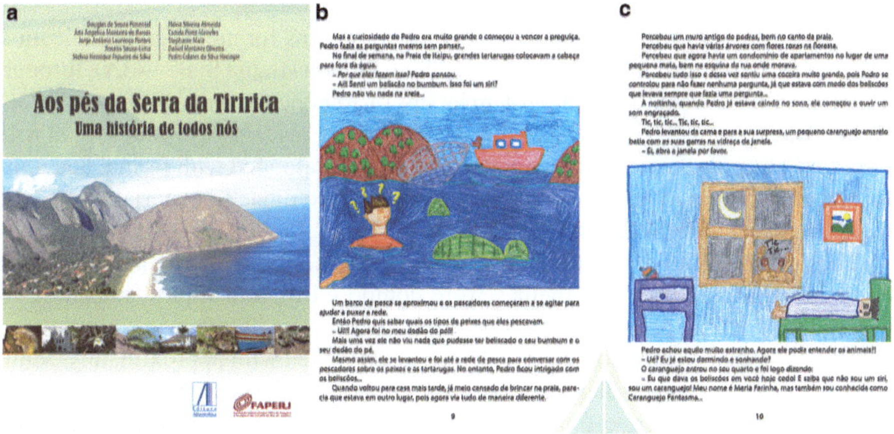

Fig. 8.2 (**a**) Book cover and (**b, c**) parts of the history elaborated to collect children's drawings for the project "Monitoring and Diagnosis of Itaipu Lagoon Health and Fisheries Resources." Two main characters were created: Pedro (**b**), which is a school student and Maria Farinha (**c**), which is an imaginary crab that brings information to the history of the book. Source: Pimentel et al. (2013)

produced highlights the local region's socio-environmental interactions across the cultural, historical, and environmental context of the Serra da Tiririca region. This also ratified the local institutions' relevance as nonformal learning spaces (Pimentel et al. 2016) and encouraged the interdisciplinary cooperation of teachers in the school.

Another example of book that can be used in different places and contexts is the Manual on Marine and Coastal Ecosystems for Educators ("Manual de Ecossistemas Marinhos e Costeiros para Educadores") (Gerling et al. 2016), developed by the BIOMAR Network (Network of Projects on Marine Biodiversity). The manual (available at: https://tinyurl.com/ycwo8rgw) presents information regarding: (1) the Brazilian oceanic and coastal ecosystems, (2) ecosystem services, (3) sustainability, (4) conservation challenges, and (5) protection and management. In the end of the book, several educational activities are presented, all of them developed by the projects that constitute the BIOMAR Network.

Also, we can mention the guide for environmental education on rocky shores (Ghilardi-Lopes et al. 2012), which presents the most common organisms of the southeast coast of Brazil and introduces basic aspects and curiosities related to their environment and taxonomic groups as well as presents a chapter about anthropogenic impacts on the sea, another on environmental education in marine environments, and one about incidents involving marine invertebrates.

Finally, the Coral-Sol Project developed a comic book to promote the awareness of the population regarding the invasion of sun coral on Brazilian waters (Andrade et al. 2013).

8.4 Radio Programs

In this section, we will describe the intervention options for media communication which were developed in the Recôncavo Baiano Region (Bahia State—BA), through a partnership between the Federal University of Recôncavo Baiano and the NGO BiomaBrasil, from 2012 to 2016, with the purpose of rescuing the educational role of radios, putting socio-environmental themes to be dialogued with formal and nonformal education publics as well as producing diverse didactic material, including digital audio files, booklet, and videos.

The project, entitled "Environmental Encyclopedia," brought together 16 themes of regional importance. The themes were composed of four topics, intending to present a concept, a problem, a curiosity, and an ecotactic (Table 8.1).

The programs were recorded in a studio as MP3 files with approximately 5 min each and were available to the local radio station. The teachers of the Basic Education (elementary till high school) of the cities received a calendar with all the scheduled programs, so that they previously knew the theme of each week. In addition, the same topic was repeated 2 or 3 times a day, always at previously defined times. So, if it was not possible to listen to the program in the classroom, it would be possible to hear it in the school counterturn, as a reinforcement or complementary activity.

Table 8.1 Examples of schedule for some themes/topics of the radio programs of the Environmental Encyclopedia project, which is the result of a partnership between the Federal University of Recôncavo Baiano and the NGO BiomaBrasil

Week theme/ Topic	Monday concept	Tuesday problem	Wednesday curiosity	Thursday ecotactic	Friday reprise
Week 01 Extractive Reserve	Extractive Reserve: A pact with nature	The tragedy of the commons	Knowing a reserve	How to conserve an Extractive Reserve	Repeat three programs
Week 02 Management Council	Management Council. Our hands at mass	Collectivism versus Individualism	Who forms our Council?	From the individual to the World	Repeat three programs
Week 03 Watershed	More than a beautiful name: Watershed	Using and abusing without care!	Knowing a friend river	The watershed as a territorial unit	Repeat three programs
Week 04 Mangroves	Mangroves: where life explodes	Destroying the chicken of the golden eggs	Oh, that! It's mangrove stuff!	Caring for sea eyelashes	Repeat three programs

Each MP3 file was derived from a text (script) of about 500 words. The 64 scripts produced were gathered in two booklets, also previously distributed to teachers to complement the actions of research and reading in the classroom. At the end of 52 weeks (one year), it was possible to produce 208 programs, constituting the Environmental Encyclopedia.

One thousand copies of the booklets were distributed (500 in Maragogipe and 500 in Sapeaçu City, BA) during the Pedagogical Planning Week in the beginning of the year. One of the scripts was adapted to produce an audiovisual which presented a dialogue between two scientists about the ecosystem topic.

Until now, the perception of the coordinators of the project is that it was successful and with large potential for replication. The main challenge was related to the consequences of the "acoustic monopoly" (Brecht 2003) which hampers the dialogue with the radio stations and makes it difficult to design initiatives in tune with formal education, even if we consider that this is a criterion for the concessions of broadcasting services (Andrelo and Kerbauy 2009).

Also, it was important to revisit the "Theory of Radio" (Brecht 2003) and the contributions of the Frankfurt School, especially about the risks in a consumer society mediated by a technological context that imposes standards, ideologies, and forms the "One-Dimensional Man" (Marcuse 1973) resulting from the extrapolation of "technological rationality". This theoretical–practical exercise allowed for reflections that pointed out new paths to this project:

- Radio is a media in which regionalization is very important (Gambaro 2009) so that content cannot be widely replicated without renewal/adaptation;
- Sharing of MP3 files directly with teachers of Basic Education should be prioritized, without excluding the possibility of their broadcasting. The structuring of a web radio accompanied by the sharing of podcasts is not advantageous in the

context of the Recôncavo Baiano. For example, 42% of the 68 million permanent private households in Brazil do not have access to the internet in 2015; and in Bahia, only 32.5% of the households had computers (IBGE 2016), with an even smaller percentage of access to the internet service.
- The radiophonic language has "educational potential," but must be converted in an educational strategy to build knowledge, including the epistemic domains (affective, cognitive, and behavioral) (Brando da Rocha et al. 2009); so, it is necessary to combine content and educational goals in order to differentiate information from education (Andrelo and Kerbauy 2009).

8.5 Games

Complex socio-environmental issues such as global climate changes and environmental degradation can be better understood if accompanied by interactive learning strategies, which are related to experiential learning (Kolb 1983), enabling the integration of apprehension with comprehension (Dieleman and Huisingh 2006) besides the direct and experiential interaction of the learner with the knowledge and, finally, the extension of all of this to a behavior change and promotion of actions that aim the improvement of socio-environmental quality (Ghilardi-Lopes et al. 2013).

One Brazilian example is a digital educational interdisciplinary and systemic gamebook (Ghilardi-Lopes 2014) on the influences of climate changes on marine and coastal environments. The game is entitled "Apicum challenge" (Desafio em Apicum, in Portuguese) and is available for free at https://tinyurl.com/mstdf3h. The game is aimed at students in junior high school and high school. The scenario of the game is a coastal fictitious city called Apicum and the main character will have to understand the causes and effects of climate changes on the coastal and marine environments of the city. Throughout the levels (Table 8.2), the participant will gain points of "Environment," "Society", and "Economy," the three pillars of sustainability.

Another example is the game "Poseidon—o Que Está Em Jogo Neste Jogo" (Bensusan 2010—available at https://tinyurl.com/k9c7w3c). This didactic material is a card game aimed at children above 6 years old which presents how difficult is the process of establishing a Marine Protected Area. The game is constituted of 64 cards representing "threats," "defenses," "fish," and four different "Brazilian Marine Protected Areas." There is also one card with the instructions of the game. Also, the game is accompanied by one book about the importance of the sea, its legends, richness, impacts, threats, and some conservation instruments, such as protected areas.

The Coral-Sol Project in Brazil also developed a game related to marine biodiversity and invasion. The idea of the game is based on the hot-potato game, in which an invasive coral skeleton is passed from hand to hand while a song is played. When the song ends, the person with the skeleton must randomly pick a card, which contains an image of a native organism. The participant must try to identify the organism and learn its characteristics. Also, the group must reflect upon the possibility of coexistence between the native species and the invasive sun-coral.

Table 8.2 Levels of the "Apicum challenge," an educational game on global climate changes and their effects on marine and coastal ecosystems, with the objectives proposed for each one

Level	Objective(s)
Introduction	Help the people who suffered with an extreme storm Talk with marine ecologists
School	Learn about marine ecosystems and global climate changes
Character's home	Make his/her home more sustainable regarding greenhouse gases emissions
Coral reef	Visit a coral reef and measure environmental variables to understand why the corals are bleached
Store	Buy diving equipment
Library	Learn about the issues dealt within the game
University	Talk to a marine researcher and get help to understand the problem with the coral reef
Laboratory	Learn about equipment to measure environmental variables
City Hall	Choose for a candidate considering the three dimensions of sustainability

Other proposals of games have been developed by the project "From Garbage to Animals: proposition of playful materials made of waste in environmental education actions" (Peixe et al. 2018), coordinated by professors Rita Inês Petrykowski Peixe and Laura Pioli Kremer from Instituto Federal de Santa Catarina (IFSC), Itajaí. The games are proposed collectively by professors and students from professional high school and vocational education and stimulate the use of waste materials originated from the industries (textile, metal mechanics, timber, and thermoplastic, among others) by a collaborative network of institutions of education, generators of waste, and resocialization (part of the games are made in the worksite of a prison). The development of these games is based on research on the production of toys, and they are used in environmental education actions, primarily aimed at elementary and high school education. The project includes the development of several games that have as prerequisites: the inclusion of social, environmental, and cultural aspects; the playful nature of the proposal; and the use of waste from the industries. Until present, three prototypes are in the final stage of elaboration, and they are listed below:

"Bichoruga" (Animal Turtle) This game is an adaptation of the game named Cinco Marias (Chuck stones or Jack stones), which was originated in Ancient Greece (Meirelles 2007). The original game consists on the development, in the participants, of spatial perception, interaction, and motor coordination through the throwing of stones upwards, and the performance of certain feats while the stones are on the air. The resignification of the game brought up concepts related to production, consumption, and waste disposal and its consequences for society. This includes the Brazilian guidelines related to solid waste, the whole problem of solid waste in the oceans, and the loss of biodiversity as a consequence of the ingestion of waste by organisms, among others. The five original parts of the game are represented by jellyfishes, named "águas-marias." These playful elements are lodged inside a bag which represents a marine turtle named "BICHORUGA," the animal turtle (Fig. 8.3a). The reinvention of this game alludes to the resemblance between the jellyfishes, which is the natural food of certain species of marine turtles, and the ingestion of plastic bags (solid waste), one of the main causes of their mortality.

All for One Planet This proposal is an adaptation of a classic action game, widely spread in Brazil. In the original game, which may include 2–4 players, it is necessary to keep some marbles in the upper part of a vertical tube, supported by a network of sticks, which cross this tube horizontally through holes on the sides of the tube. Each player takes away one stick at a time so that marbles will not fall to the base of the tube. In the adaptation proposed by the project, the base of the tube is represented graphically by nine sections that correspond to the nine planetary boundaries (Rockstrom et al. 2009), which represent nine large environmental processes crucial to define safe critical boundaries for sustaining the resilience of ecosystems, namely: climate change, ocean acidification, stratospheric ozone, global phosphorus and nitrogen cycles, atmospheric aerosol loading, freshwater use, land use change, biodiversity loss, and chemical pollution (see more in Chap. 1). The players, in a collaborative manner, must balance their actions (represented by the withdrawal of the sticks), aiming to avoid crossing over the safe critical boundaries for the conservation of the resilience of the Earth system. The crossing of a boundary is a result of the accumulation of marbles transferred to each of the sections of the nine planetary limits chart in the base of the tube, which has a limit of five marbles. The player who will remain in the game is the one who will avoid the fall of the fifth marble in any of the sections (Fig. 8.3b).

Game of the Marine Bioinvasion This is an adaptation of a Brazilian two-player board game of indigenous origin, named "Jogo da Onça" (game of the jaguar) (Meirelles 2007), in which one player represents 14 dogs which have the challenge of cornering their opponent, a jaguar (a big Brazilian feline), represented by the second player. The defense of the jaguar is the possibility of capturing the dogs through movements in the board. In the adaptation of the game to the marine context, the conflict between the jaguar and the dogs was modified to represent concepts of bioinvasion. In this game (Fig. 8.3c), the jaguar original character was replaced by the invasive species commonly known as sun-coral, introduced unintentionally in the Brazilian Coast (*Tubastraea coccinea* and *Tubastraea tagusensis*). The dogs are represented by the endemic species of coral from the Brazilian coast (*Mussismilia* spp.). In the context of the game, the endemic coral species must compete with the exotic ones to limit their distribution. The game ends when the native corals corner the exotic species in the board, which represents the biotic resistance of the communities, the native communities' capacity to keep the exotic species from colonizing and establishing themselves in the environment where they were introduced.

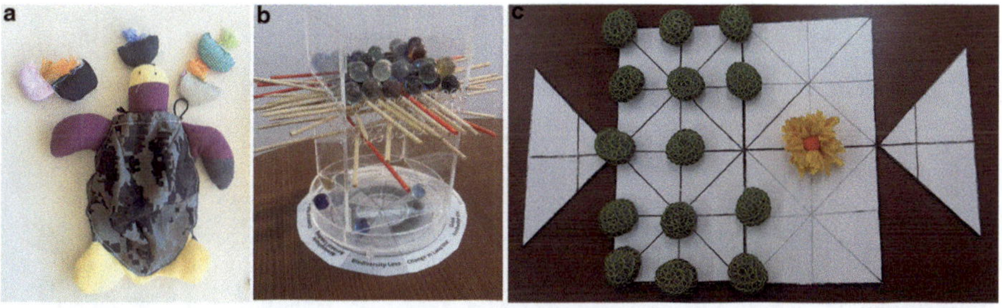

Fig. 8.3 Games developed by the Project "From Animal to Garbage.": (**a**) Bichoruga, (**b**) All for one planet, and (**c**) Game of the marine bioinvasion. Photos by: Laura P. Kremer

8.6 Citizen Science Protocols

Active learning strategies may occur in natural environments and involve the scientific education of learners. Citizen science is a process that involves the general public in scientific projects (Bonney et al. 2009) and can be a powerful tool for marine and coastal environmental education (Ghilardi-Lopes 2015) (see also Chaps. 4 and 7). Standardized protocols and low-cost materials are a key to the success of citizen science initiatives. In Brazil, for example, the projects "Meros do Brasil" and "Mantas do Brasil," both related to fish species, developed educational materials (Gerhardinger et al. 2010; Kodja 2015). At the Federal University of ABC (São Paulo State), two citizen science protocols related to coastal and marine environments were developed: one is related to the use of euglenas as bioindicators to monitor beach quality and the other is related to the long-term monitoring of benthic organisms in marine protected areas (Kawabe 2017). Both protocols can be accessed at http://professor.ufabc.edu.br/~natalia.lopes/cienciacidada/. In South America, one example of successful citizen science initiative, related to microplastics in beaches, was developed in Chile (Hidalgo-Ruz and Thiel 2013), and educational materials were also produced (available at http://www.cientificosdelabasura.cl/material.php).

8.7 Materials for Exhibitions

Exhibits are widely used in formal education or as a free choice learning experience (Ballantyne and Packer 2005), being important vehicles for building knowledge, transforming values, and raising public awareness when carried out within critical strands. For this, they can use thematic spaces for the promotion of debates, group dynamics, and even mobilization (campaigns or community meetings, for example) for the conservation of marine and coastal biodiversity. They also make it possible

Fig. 8.4 (**a**) Itinerant truck used by the Ilhas do Rio Project and (**b**) some instructional materials used in the exhibitions. Photos by: Camila P. Meireles

to bring the public closer to the marine or coastal environments addressed, which may often be inaccessible.

In Rio de Janeiro State, the Ilhas do Rio Project, organized by the Instituto Mar Adentro, set up an itinerant truck exhibition (Fig. 8.4) on the Cagarras Islands Natural Monument. The trunk of the truck was built to carry a TV, a library for the didactic-scientific zoological collection of the Teaching Assistance Section of the National Museum of Rio de Janeiro, magnifying glasses, research equipment for manipulation by the public, sensory boxes for touching parts of organisms (e.g., shells, carapaces, and skeletons), explanatory panels, photographic shows, diving mannequins, living and herbarium specimens of plants as well as thematic tents for children. Booklets and brochures were designed and distributed to the visitors to deepen their knowledge regarding the exhibition. The initiative reached thousands of people and allowed residents and tourists in the city to be surprised by the biodiversity of the islands so close to the coast, but so unknown, arousing their notion of belonging.

8.8 Final Considerations

Didactic materials are fundamental for the development of EE activities, since they can help learners and educators promoting environmental awareness and achieving educational goals. They can vary in accordance with the context, goals, and financial resources of each EE initiative and their development can involve people with diverse instructional backgrounds and abilities, which can be positive in educational terms, especially when the creation process is collaborative. Some examples of instructional/didactic materials developed for CMEE activities in Brazil and how they can contribute to enhance teaching–learning process was presented in the present chapter. It is possible to notice that although there is a great effort of different institutions for the development of these materials, their sharing and evaluation is still incipient. According to Simmons (1996), some key guidelines and indicators can be taken into account for the selection or development of instructional/didactic materials: (1) fairness and accuracy, (2) depth, (3) emphasis on skills building, (4) action orientation, (5) instructional soundness, and (6) (re)usability. Those same indicators could be used for their evaluation and they should be prioritized if we aim at high-quality and effective environmental education (see more about evaluation in Chap. 5).

References

Afonso MCL, Eirão TG, Melo JHM, Assunção JS, Leite SV (2011) Banco Internacional de Objetos Educacionais (BIOE): tratamento da informação em um repositório educacional digital. Perspect Cienc Inf 16:148–158. https://doi.org/10.1590/S1413-99362011000300009

Andrelo R, Kerbauy MT (2009) Gênero educativo no rádio: parâmetros para a elaboração de programas voltados à educação. Intercom Rev Bras Cienc Comunicação 32:147–164

Andrade A, Meireles C, Ribeiro F, Creed JC (2013) A invasão do Coral-Sol - uma ameaça aos nossos mares. Instituto Brasileiro de Biodiversidade Marinha, Rio de Janeiro

Ballantyne R, Packer J (2005) Promoting environmentally sustainable attitudes and behaviour through free-choice learning experiences: what is the state of the game? Environ Educ Res 11:281–295. https://doi.org/10.1080/13504620500081145

Barcellos M, Maia S, Meireles C, Pimentel DS (2015) Elaboração da trilha interpretativa do Morro das Andorinhas: Uma proposta de Educação Ambiental no Parque Estadual da Serra da Tiririca. In: Vallejo LR, Pimentel DS, Montezuma RC (eds) Uso Público em Unidades de Conservação. Planejamento, turismo, lazer e impactos, pp 391–400

Bensusan N (2010) Poseidon - O Que Está Em Jogo Neste Jogo. Ministério do Meio Ambiente, Brasília

Berchez F, Ghilardi N, Robim MJ et al (2007) Projeto Trilha Subaquática: sugestão de diretrizes para a criação de modelos de educação ambiental em unidades de conservação ligadas a ecossistemas marinhos. OLAM Cienc Tecnol 7:181–209

Bonney R, Cooper CB, Dickinson J et al (2009) Citizen science: a developing tool for expanding science knowledge and scientific literacy. Bioscience 59:977–984. https://doi.org/10.1525/bio.2009.59.11.9

Brando da Rocha F, Plácido TP, Caldeira A (2009) Proposta didática para o ensino de ecologia. Enseñanza Cienc 8:1789–1794

Brasil (1998) Parâmetros curriculares nacionais: Ciências Naturais. MEC/SEF, Brasília

Brecht B (2003) Teorias de la Radio (1927–1932). Rev Econ Política Tecnol Inform Comun 5(2):5–16

Chiappe Laverde A, Segovia Cifuentes Y, Rincón Rodríguez HY (2007) Toward an instructional design model based on learning objects. Educ Technol Res Dev 55(6):671–681 https://doi.org/10.1007/s11423-007-9059-0

Dieleman H, Huisingh D (2006) Games by which to learn and teach about sustainable development: exploring the relevance of games and experiential learning for sustainability. J Clean Prod 14:837–847. https://doi.org/10.1016/j.jclepro.2005.11.031

Fernandes R, Pimentel DS, Ferreira MS (2011) A Educação Ambiental como mediadora na diluição de fronteiras institucionais. In: Seabra G, Mendonça I (eds) Educação Ambiental: Responsabilidade para a conservação da sociobiodiversidade. Editora Universitária da UFPB, João Pessoa, pp 1461–1469

Gambaro DO (2009) Uso das novas TICs pelas emissoras de rádio: uma análise dos casos paulistanos e o referencial de Bernard Miège. In: XXXII CONGRESSO BRASILEIRO DE CIÊNCIAS DA COMUNICAÇÃO, Curitiba, PR, 2009. INTERCOM - Sociedade Brasileira de Estudos Interdisciplinares da Comunicação

Gerhardinger LC, Borgonha M, Bertoncicni AA (2010) Memória do mar: biodiversidade, conservação e cultura no litoral brasileiro. Ecomares, Florianópolis

Gerling C, Ranieri C, Fernandes L et al (Orgs) (2016) Manual de ecossistemas marinhos e costeiros para educadores. Editora Comunnicar, Santos. http://www.icmbio.gov.br/portal/images/stories/ManualEcossistemasMarinhoseCosteiros3.pdf

Ghilardi NP, Berchez F (2010) Projeto Trilha Subaquática - modelos de educação ambiental marinha para o grande público. In: Pedrini AG (ed) Educação ambiental marinha e costeira no Brasil. EdUERJ, Rio de Janeiro, pp 71–91

Ghilardi-Lopes NP (2014) Desafio em Apicum. UFABC, Santo André

Ghilardi-Lopes NP (2015) Citizen science combined with environmental education can be a powerful tool for coastal-marine management. J Coast Zone Manag 18(407):27. https://doi.org/10.4172/2473-3350.1000407

Ghilardi-Lopes NP, Hadel VF, Berchez F (2012) Guia para Educação Ambiental em Costões Rochosos. Artmed, Porto Alegre

Ghilardi-Lopes NP, Braga JC, Silva L et al (2013) Environmental education through an online game about global environmental changes and their effects on coastal and marine ecosystems. In: Proceedings of SBGames 2013, São Paulo, 2013. Art & Design Track - Full Papers. Sociedade Brasileira de Computação, pp 469–474

Giraldo M, D'amico D (2016) When actions challenge theories: the tactical performances of Sue Austin. Perform Matter 2(1):38–50

Gokhale AA (1995) Collaborative learning enhances critical thinking. J Technol Educ 7(1). https://doi.org/10.21061/jte.v7i1.a.2

Hidalgo-Ruz V, Thiel M (2013) Distribution and abundance of small plastic debris on beaches in the SE Pacific (Chile): a study supported by a citizen science project. Mar Environ Res 87-88:12–18. https://doi.org/10.1016/j.marenvres.2013.02.015

IBGE (2016) Pesquisa Nacional por Amostra de Domicílios. Acesso à internet e à televisão e posse de telefone móvel celular para uso pessoal. Instituto Brasileiro de Geografia e Estatística, Rio de Janeiro

Karyn Benton MS (2011) Developing a multi-sensory outdoor education program. Alexandria 4(4):177–179

Kawabe LA (2017) Monitorando as UCs marinhas: guia para foto-identificação de biótopos e monitoramento participativo de unidades de conservação (UCs) marinhas. Ciência cidadã, vol 5, 1 edn. UFABC, Santo André

Kodja G (2015) Estudo e conservação das raias mantas. Instituto Laje Viva, Santos

Kolb D (1983) Experiential learning, experiences as the source of learning and development. Prentice Hall, Englewood Cliffs, NJ

Koppi T, Bogle L, Bogle M (2005) Learning objects, repositories, sharing and reusability. Open Learning: J Open Dist e-Learn 20(1):83–91. https://doi.org/10.1080/0268051042000322113

Marcuse H (1973) A ideologia da sociedade industrial: o homem unidimensional, 4th ed. Zahar Editores, Rio de Janeiro

Meireles CP, Ribeiro FV, Andrade AC et al (2013) Implantação da Primeira Trilha Interpretativa Terrestre e Subaquática em Área de Proteção Ambiental na Ilha Grande (Angra dos Reis, RJ). In: Costa NMC, Costa VC, Mello FAP (eds) II Congresso Nacional de planejamento e manejo de trilhas. Rede Sirius, Rio de Janeiro, pp 863–881

Meireles CP, Creed JC, Pimentel DS (2016) A Educação Ambiental no controle da bioinvasão marinha por Coral-Sol (*Tubastraea* spp. Anthozoa, Dendrophylliidae) em Angra dos Reis (RJ, Brasil). Ambient Sustentable 2:1–21. https://doi.org/10.17979/ams.2015.02.020.1612

Meirelles R (2007) Giramundo e outros brinquedos e brincadeiras dos meninos do Brasil. Editora Terceiro Nome, São Paulo

Padrón C, Dodero J, Díaz P, Aedo I (2005) The collaborative development of didactic materials. Comput Sci Inf Syst 2(2):1–21

Pedrini AG, Brotto DS, Lopes MC, Messas TP (2011) Gestão de Áreas Protegidas com Educação Ambiental Emancipatória pelo Ecoturismo Marinho: A Proposta do Projeto EcoTurismar. OLAM Cienc Tecnol 10:6–81

Peixe RIP, Kremer LP, Galindo L, Henning TEA, Novais MC, Steinbraenner L (2018) A hora e a vez dos resíduos: quando o lixo pode virar bicho. In: Congresso Internacional e Workshop Design & Materiais, Joinville, 2018. UNIVILLE - Universidade da Região de Joinville

Pimentel DS, Monteiro de Barros AA, Motta ES et al (2004) Society and conservation units: the challenge of Serra da Tiririca State Park, Niterói/Maricá, RJ, Brazil. In: Leal Filho W, Littledyke M (eds) International perspectives in environmental education, Environmental education, communication and sustainability, vol 16. Peter Lang, Frankfurt, pp 119–128

Pimentel DS, Barros AAM, Pontes JAL et al (2013) Aos pés da Serra da Tiririca: Uma história de todos nós vol 1. Alternativa, Niterói

Pimentel DS, Meireles CP, Maia SM, Barcellos MM (2016) Interpretative trails to enable an environmental education process in a Brazilian park. In: Vasiljević Đ (ed) 8 monitoring and management of visitors, Novi Sad, Serbia, 2016. Novi Sad University, pp 393–396

Pimentel DS, Barcellos MM, Meireles CP, Oliveira ML, Santos VP (2017) Trilhas interpretativas como estratégia de educação ambiental em unidades de conservação. In: Santori RC, Santos, MG, Santos, MCF. (eds) da Célula ao Ambiente: Propostas para o ensino de Ciências e Biologia. UERJ/FFP, Rio de Janeiro, pp 179–201

Remillard JT, Heck DJ (2014) Conceptualizing the enacted curriculum in mathematics education. In: Thompson DR, Usiskin Z (eds) Enacted mathematics curriculum: a conceptual framework and research needs. Information Age Publishing, Charlotte, NC, pp 121–148

Rhormens MS, Pedrini AG, Ghilardi-Lopes NP (2017) Implementation feasibility of a marine ecotourism product on the reef environments of the marine protected areas of Tinharé and Boipeba Islands (Cairu, Bahia, Brazil). Ocean Coast Manag 139:1–11. https://doi.org/10.1016/j.ocecoaman.2017.01.022

Rockstrom J, Steffen W, Noone K et al (2009) A safe operating space for humanity. Nature 461:472–475. https://doi.org/10.1038/461472a

Seven SM (1980) Environmental interpretation for the visually impaired. J Environ Educ 11:37–40. https://doi.org/10.1080/00958964.1980.9941389

Simmons D (1996) Environmental education materials: guidelines for excellence. NAAEE, Washington, DC

Watson AP (1982) The design of self-guiding, interpretative, nature trails accessible to the disabled. Arboric J 6:287–293

Wiley DA (2002) Connecting learning objects to instructional design theory: a definition, a metaphor, and a taxonomy. In: Wiley DA (ed) The instructional use of learning objects. AECT, Bloomington, pp 3–23

Index

N. P. Ghilardi-Lopes, F. A. S. Berchez (eds.), *Coastal and Marine Environmental Education*, Brazilian Marine Biodiversity,
https://doi.org/10.1007/978-3-030-05138-9

MIX
Papier aus verantwortungsvollen Quellen
Paper from responsible sources
FSC® C105338
FSC
www.fsc.org

If you have any concerns about our products,
you can contact us on
ProductSafety@springernature.com

In case Publisher is established outside the EU,
the EU authorized representative is:
Springer Nature Customer Service Center GmbH
Europaplatz 3, 69115 Heidelberg, Germany

Printed by Libri Plureos GmbH
in Hamburg, Germany